INTRODUCING
Mind and Brain

Angus Gellatly • Oscar Zarate

Edited by Richard Appignanesi

Icon Books UK Totem Books USA

This edition published in the UK in 2005 by Icon Books Ltd., The Old Dairy, Brook Road, Thriplow, Royston SG8 7RG
email: info@iconbooks.co.uk
www.iconbooks.co.uk

Sold in the UK, Europe, South Africa and Asia by Faber and Faber Ltd., 3 Queen Square, London WC1N 3AU or their agents

Distributed in the UK, Europe, South Africa and Asia by TBS Ltd., Frating Distribution Centre, Colchester Road, Frating Green, Colchester CO7 7DW

This edition published in Australia in 2005 by Allen and Unwin Pty. Ltd., PO Box 8500, 83 Alexander Street, Crows Nest, NSW 2065

Previously published in the UK and Australia in 1998 under the title *Mind & Brain for Beginners* and in 1999 under the current title

Reprinted 1998, 2000, 2001, 2003

This edition published in the USA in 2005 by Totem Books
Inquiries to: Icon Books Ltd., The Old Dairy, Brook Road, Thriplow, Royston SG8 7RG, UK

Distributed to the trade in the USA by National Book Network Inc., 4720 Boston Way, Lanham, Maryland 20706

Distributed in Canada by Penguin Books Canada, 10 Alcorn Avenue, Suite 300, Toronto, Ontario M4V 3B2

ISBN 1 84046 638 3

Originating editor: Richard Appignanesi

Printed and bound in Singapore by Tien Wah Press Ltd.

INTRODUCING

'Often imitated ... seldom equalled'
John Gribbin

'Clever and witty'
Guardian

'Most recommended'
Time Out

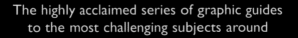

The highly acclaimed series of graphic guides to the most challenging subjects around

introducingbooks.com

INTRODUCING is the brilliant series of illustrated guides to the big thinkers and subjects from the history of ideas.

Written by experienced academics in collaboration with leading graphic artists, **INTRODUCING** is famous for being one of the best – and most enjoyable – ways to get your head around the most challenging ideas there are.

Anthropology
Barthes
Buddha
Chaos
Chomsky
Consciousness
Critical Theory
Cultural Studies
Darwin
Derrida
Descartes
Einstein
Empiricism
The Enlightenment
Ethics
Evolution
Evolutionary Psychology
Existentialism
Foucault
Fractal Geometry
Freud
Genetics
Heidegger
Hinduism
Islam
Jung
Kant
Lacan
Linguistics
Logic
Marx
Marxism
Mathematics
Media Studies
Melanie Klein
Mind and Brain
Modernism
Nietzsche
Philosophy
Plato
Postmodernism
Psychology
Quantum Theory
Romanticism
Sartre
Semiotics
Sociology
Stephen Hawking
Time
Wittgenstein

Each book: UK £9.99 * Can $17.00 * USA $12.95 * 210 x 140 paperback * 176 pp

WIN THE ENTIRE INTRODUCING SERIES!

Send your name, address and email to: introducing@iconbooks.co.uk

You will be signed up to join the Icon e-newsletter and, with each newsletter sent, ten lucky subscribers will win every book in the series!

introducingbooks.com

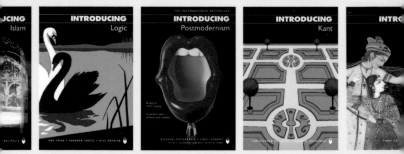

This book is about a biological organ, **the brain**, and what it does, **the mind**.

As with all body parts, evolution has adapted brains to suit particular environments and ways of life. If the brain has evolved, and is the vehicle of the mind, does it follow that the mind has also evolved? The answer to this question must be both "Yes" and "No". The brain and "biological mind" of primates **evolved** for life in the jungle or out on the savannah. They are adapted to solve the particular problems of finding food and shelter, of reproducing and caring for young.

However, in addition to being an **evolved** "biological mind", the human mind is also a "cultural mind" **socialized** in how to solve a host of "unnatural" problems thrown up by the invention of music-making and reading, painting, computer-programming and voting in elections. The cultural mind is reflexive – it reflects upon itself. To an extent, the mind is how we talk and think about it.

Mind and Brain: a Brief History

Human beings have known about the brain for a long time without being at all clear exactly **what it is for**. The large number of early **hominid skulls** which show signs of deliberate damage suggest that by three million years ago our predecessors had at least worked out that the brain is **a vital organ**.

The opening scene from Stanley Kubrick's sci-fi film classic, *2001* (1968), depicts our hominid ancestors discovering homicide.

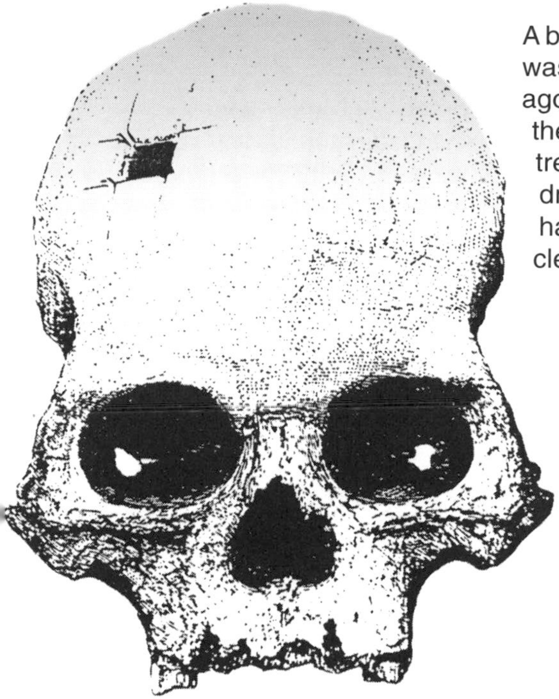

A better-intentioned knowledge was in evidence by 10,000 years ago. Neolithic skulls from around the world exhibit holes trepanned – that is, scraped or drilled – in them. The holes have **smooth edges** and show clear **signs of healing**.

IT IS LIKELY THAT TREPANNING WAS A TREATMENT FOR HEADACHES, CONVULSIONS AND INSANITY - OR "SPIRIT POSSESSION".

Paul Broca (1824-80)

Trepanning was practised until relatively recent times in Europe, and continues in many cultures. Theoretical arguments for the modern technique of **electro-convulsive therapy** (**ECT**) are scarcely stronger than those for trepanning.

When Neolithic "doctors" trepanned a "patient", did they believe they were treating the body, the mind, the spirit or the soul? We can never know. But they probably would not have recognized these distinctions.

Inventing the Mind

The epic poems of Homer in the 8th century B.C. are Europe's earliest substantial pieces of writing. The **Iliad** recounts the Siege of Troy and the **Odyssey** tells of the journey home from Troy of the hero Odysseus (Ulysses to the Romans).

Amazingly, these works scarcely refer to what we would call "The Mind". Homer's vocabulary does not include mental terms such as "think", "decide", "believe", "doubt" or "desire". The characters in the stories do not **decide** to do anything. They have no **free will**.

WE ACT ONLY BECAUSE INSTRUCTED BY VOICES OR DRIVEN BY AN INNER TENSION.
OR COERCED IN SOME WAY BY A GOD.

Where we would refer to thinking or pondering, Homer's people refer to speaking to or hearing from their own organs: "I told my heart", or "my heart told me". Feelings and emotions are also described in this half-strange, half-familiar manner. Feelings are always located in some part of the body, often the midriff. A sharp intake of breath, the palpitating of the heart, or the uttering of cries **is** a feeling. A feeling is not some inner thing separate from its bodily manifestation.

The **Iliad** and **Odyssey** are written versions of "songs" originally sung by non-literate bards that expressed the beliefs and ideas of their oral culture.

WE INVENTED THE MIND AS OUR ORAL CULTURE GRADUALLY TRANSFORMED INTO A LITERATE CULTURE.

People in oral cultures do not explicitly recognize the difference between a thought and the words which express it. What you say is what you intend. Your word (not your signature) is your bond. Speech is gone the moment it is uttered. Written records, by contrast, stay fixed. You can study them at leisure. This encourages a distinction between the persisting symbols on the page and the ideas they represent. “Literal” meaning gets consistently discriminated from “intended” meaning (as in the “letter” and the “spirit” of the law).

IDEAS AND THE WORDS WHICH EXPRESS THEM ARE NO LONGER ONE AND THE SAME.
WRITING AND SPEECH ARE NOW ACTIONS EXPRESSING PRE-EXISTING THOUGHTS.

The **ratio** of rational thought branches off from the **oratio** of speech to become a separate concept. People’s actions express their thoughts and decisions they have made.

Literacy, it is argued, drives a wedge between two worlds. One is the world we hear and see, the world of talk and action. The other is the imperceptible *mental* world of thoughts, intentions and desires. Just as talk and action take place within the physical world, so literate Greeks at the time of Plato and Aristotle created a space in which to house thoughts, intentions and desires. This metaphorical space was first called the **psyche**, but now is known as the **mind**.

What is the Mind?

We can see that this question has no simple answer. Efforts to understand the relationship between brain and behaviour, or mind and brain, are really investigations of what these words ought to mean. Some brain functions – for example, control of body temperature – happen completely unconsciously. Other functions are usually unconscious but not always: for example, breathing, except when you **voluntarily** hold your breath. These might be called bodily rather than mental functions, but the distinction is not sharp.

WHEN YOU RECOGNIZE AN OBJECT, YOU BECOME CONSCIOUS OF WHAT IT IS - A BOOK, SAY. BUT YOU ARE NOT CONSCIOUS OF *HOW* YOU RECOGNIZE IT.
IN CONSCIOUSLY RECALLING SOMEONE'S NAME YOU ARE NOT OFTEN AWARE OF THE PROCESSES BY WHICH YOU RETRIEVED IT.

So perhaps recognition and recall could be thought of as bodily processes, the results of which can (sometimes) become conscious.

Although we cannot say definitely what the mind **is**, we have ideas about what it **does**. The mind allows us to **see** the world and **act** voluntarily on it. Seeing, hearing, touching and all other sensations take place in the mind. So does the experience of emotions.

MOVEMENT (OFTEN CALLED MOTOR ACTION), THINKING, REMEMBERING AND PLANNING SEEM TO ISSUE FROM THE MIND.
MIND ALSO INCLUDES THE SENSE OF SELF AND THE SENSE OF FREE WILL.

The Greeks gave us a mentalistic psychology full of words like feel, think, want and decide. This became our common sense or folk psychology. But is it adequate for present day purposes? How well do the metaphors of **mind** and **self** map onto our knowledge of how the brain works? These questions lie at the heart of this book.

Meet the Brain

The average human brain weighs around three pounds or 1.4 kilos. Its most obvious features are the **left and right hemispheres** (LH and RH), which enclose most other (sub-cortical) parts, and the walnut-shaped **cerebellum** (little brain) at the rear where the spinal cord emerges. The surface of the hemispheres is **cortical tissue** (from the Latin, *cortex*, "bark") which is crumpled or **convoluted**. The convolutions increase the cortical surface area available within the confines of the skull.

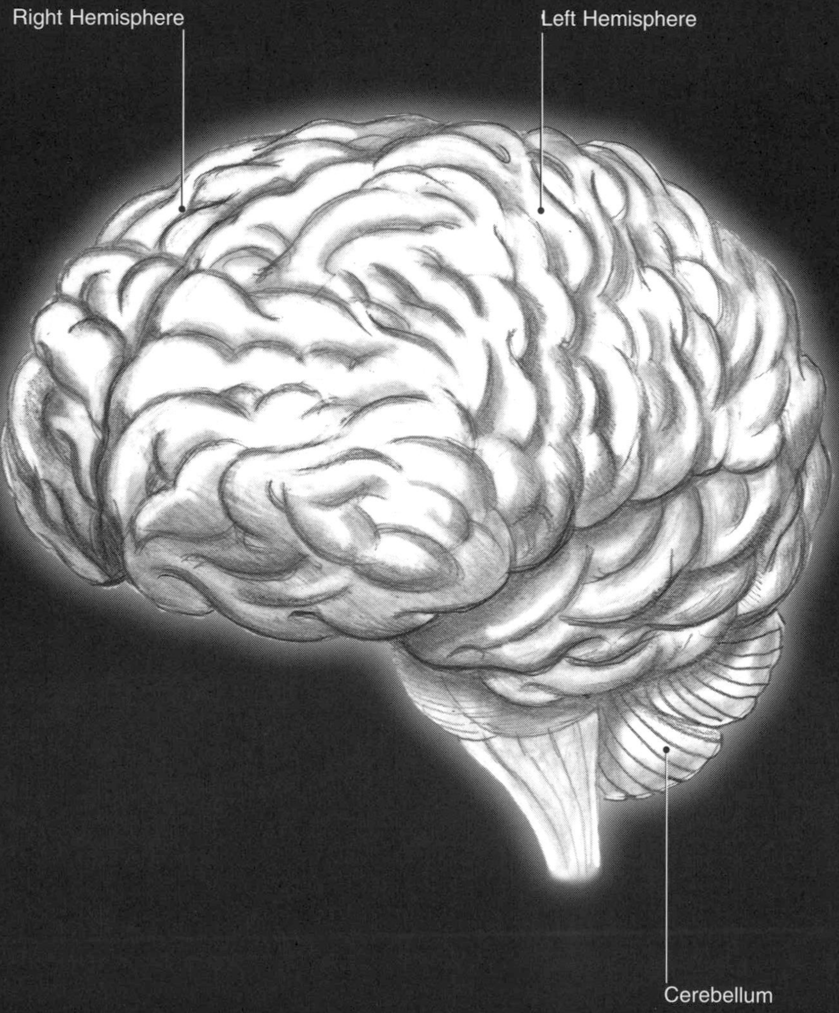

In many ancient languages, the word for **brain** and **bone marrow** was the same. The ancient Greeks and the Chinese thought both of these grew from **semen**.

Egyptians of The Middle Kingdom (c. 2040-1786 B.C.) were so unimpressed by the brain that they did not preserve it with the rest of the body, as they did the heart, lungs, liver, and kidneys.

Matter or Spirit?

The Greek physician **Hippocrates** (c. 460-377 B.C.) rejected the idea that gods and spirits cause physical and mental ill-health. He gave a purely materialist account of body and mind.

Plato (429-347 B.C.) did not accept all of this materialist humoral theory. He believed in a **Soul** with three parts.

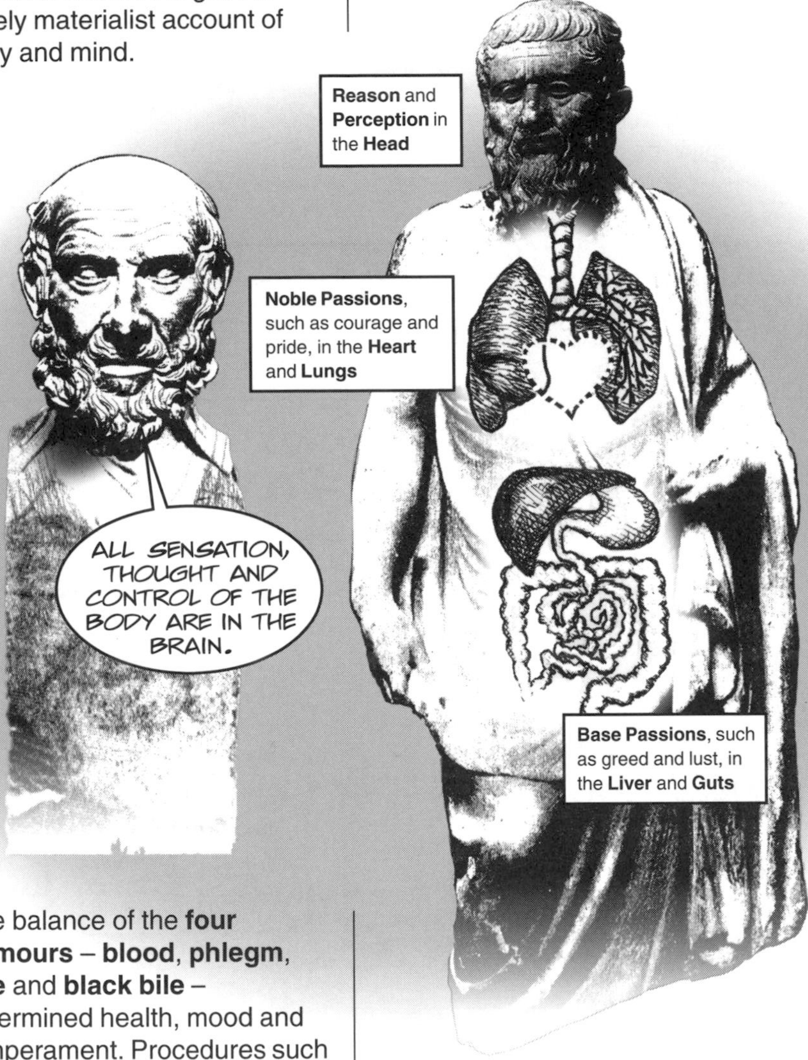

The balance of the **four humours** – **blood**, **phlegm**, **bile** and **black bile** – determined health, mood and temperament. Procedures such as bleeding, starving or purging were used to treat harmful imbalances.

The first part of the soul was **immortal**, but the second and third were **perishable**.

Aristotle (384-322 B.C.) knew that **touching** the brain does not cause any sensation. He judged that the **heart** must be where **sensations** happened.

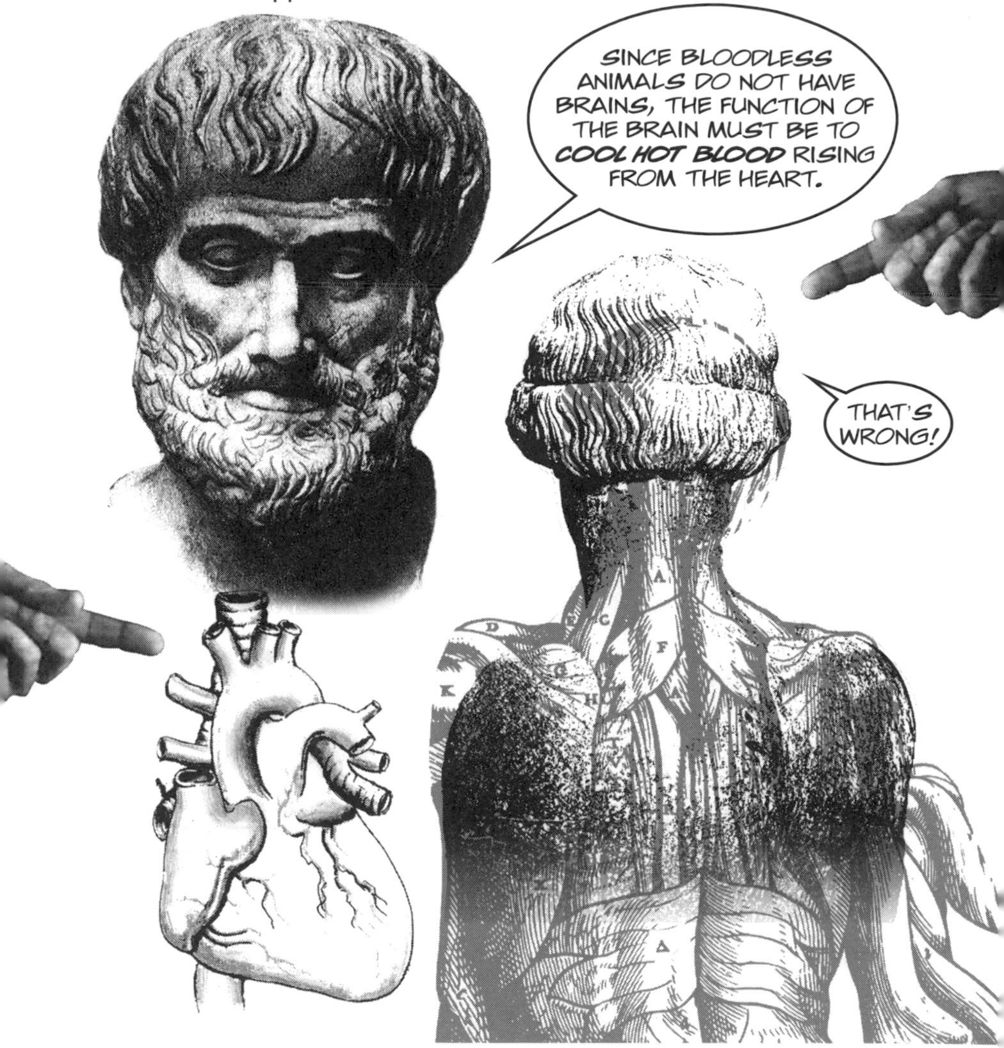

SINCE BLOODLESS ANIMALS DO NOT HAVE BRAINS, THE FUNCTION OF THE BRAIN MUST BE TO ***COOL HOT BLOOD*** RISING FROM THE HEART.
THAT'S WRONG!

Galen (129-c. 199 A.D.), a Greek physician in Roman times, relied on animal dissection, experiments, clinical practice and perhaps observations of wounded gladiators. He concluded that the brain is the organ of sensation and voluntary movement.

Debate over the **brain hypothesis** versus the **cardiac hypothesis** continued up to the Middle Ages and beyond.

Pioneer Map-Makers

The great age of European map-making and navigation began in the Renaissance. Maps were being drawn not only of "new worlds" across the seas, but up there in the skies by astronomers like **Nicholas Copernicus** (1473-1543) and **Galileo Galilei** (1564-1642), and also *inside* the body by pioneer anatomists, **Leonardo da Vinci** (1452-1519), **Andreas Vesalius** (1514-64) and others.

Stars
Saturn
Mercury
Sun
IN EVERY DIRECTION, THERE IS NEW KNOWLEDGE.

The Mind of the Gaps

From early Greek times, supporters of the brain hypothesis believed that the **soul** and **mental faculties** were located not in the tissue of the brain but in the internal cavities known as the **ventricles**.

Vesalius taught that inhaled air and **vital spirits** rising from the heart came together in the ventricles and were transformed into **animal spirits**. These were distributed through hollow channels to the organs of sensation and motion. This was an early approximation to a chemical theory of how nerves work.

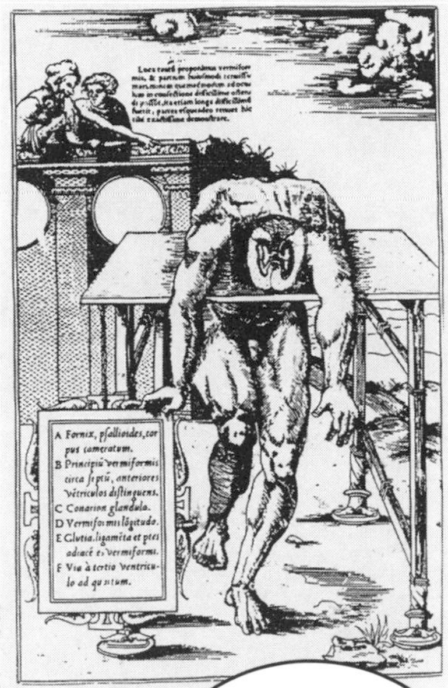

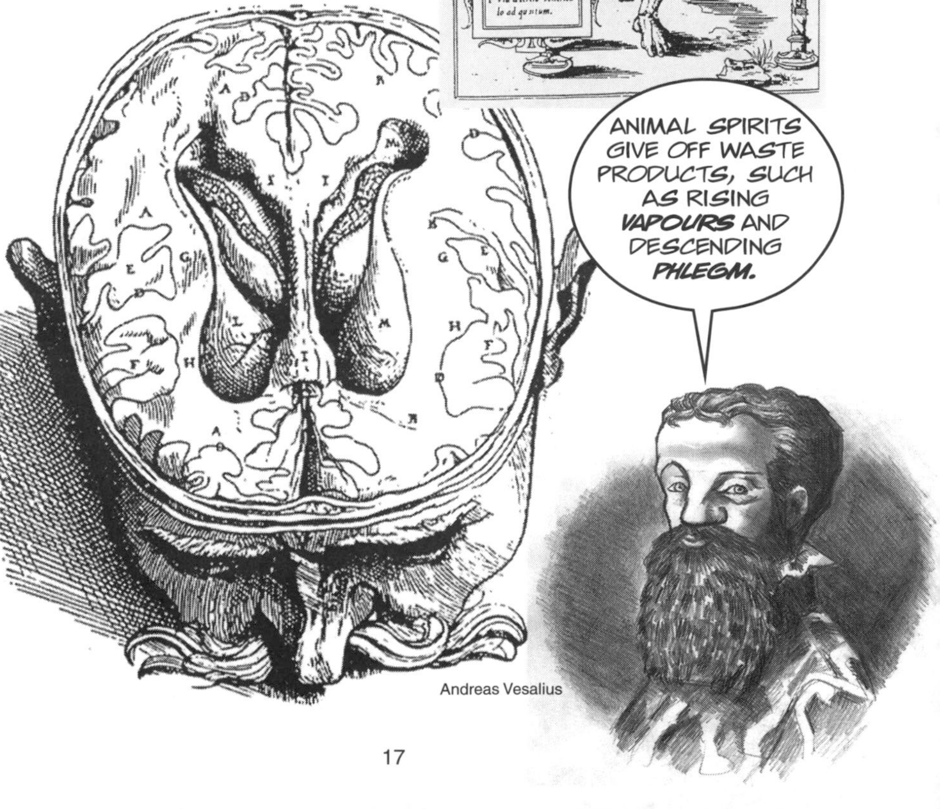

Andreas Vesalius

Ventricles, Tissues and the Mind

There were arguments over just how many ventricles the brain had. Different **functions** – such as memory, thought, judgement and reason – were said to be **localized** in different ventricles. This went on until the arrival of **Franciscus de la Boë** (known as **Sylvius**, 1614-72) and **Thomas Willis** (1621-75).

For the philosopher **René Descartes** (1596-1650), there was a total split between the conscious mind/soul and the body.

A Fish Called Wondercure

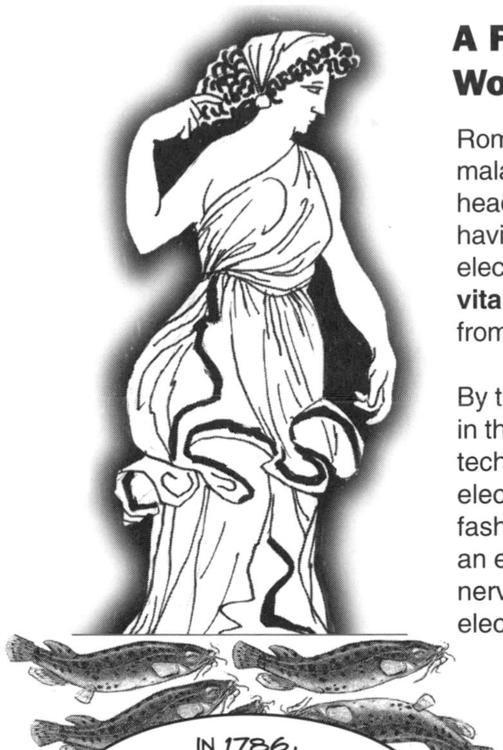

Roman physicians treated a range of maladies, including paralysis, headaches, arthritis and gout, by having their patients stand on electric fish. It was thought some **vital** or **life force** was transmitted from fish to foot.

By the mid-18th century, advances in the physics of electricity and the technology of generators brought electrical therapy once more into fashion. The brain was thought to be an electric generator, with the nerves as wires down which electrical fluid flowed.

IN 1786, I DISCOVERED THAT ELECTRICAL STIMULATION OF THE NERVES IN A FROG'S LEG CAUSED MUSCLE CONTRACTIONS.

Luigi Galvani's (1737-98) discovery laid the foundation for modern ideas of nerve conduction.

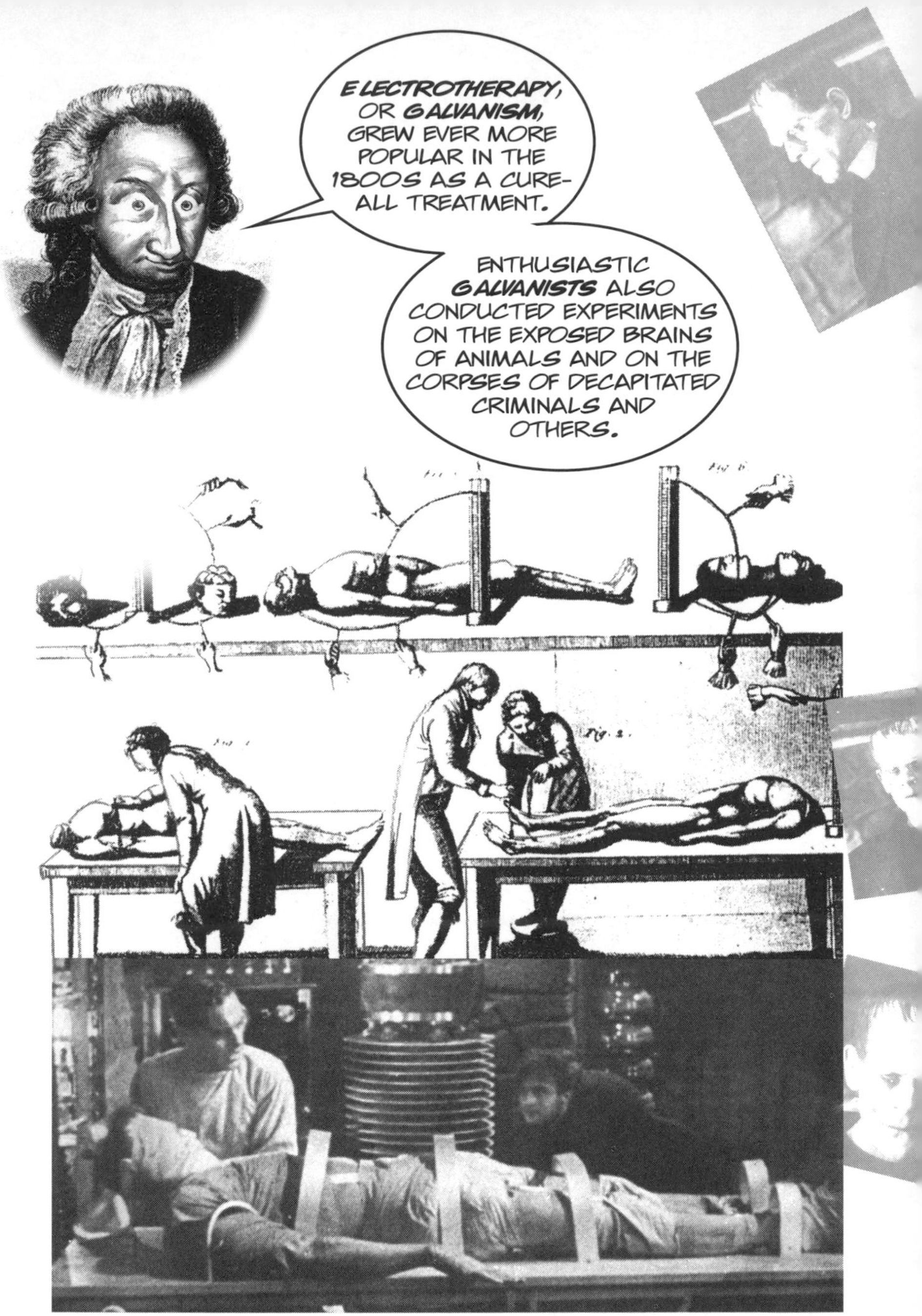

In our "surgical culture", it is easy to forget the fear and revulsion inspired by this research. But it was given expression by **Mary Shelley** (1797-1851) in her novel **Frankenstein** in 1818.

Bumps on the Head

The early 19th century also saw the development of **phrenology** by **Franz Gall** (1758-1828) and **Johann Spurzheim** (1776-1832). Both were skilled neuro-anatomists and fervently believed two things.

THE BRAIN IS THE ORGAN OF THE MIND.
DIFFERENT MENTAL AND MORAL FACULTIES ARE LOCALIZED IN PARTICULAR CORTICAL REGIONS.

Unfortunately, they also believed that the degree to which an individual possessed a faculty, such as "memory" or "love of offspring", depended on the size of the relevant brain area. This, in turn, would be reflected in the shape of the skull over that area. An affectionate parent should have a bump in the appropriate place. The idea spread that personality could be analyzed by examining the skull. Going to the phrenologist to "have your bumps felt" became as fashionable as going to your analyst would become in the 20th century. But no two phrenologists ever agreed on exactly what mental faculties there were, nor on how they were positioned over the skull.

The Beginning of Localization

Marie-Jean-Pierre Flourens (1794-1867), a strict disciple of Descartes, led the reaction against phrenology. He believed in a unified mind or soul which could not be analyzed into separate parts. Flourens studied the effects of galvanic stimulation and focal lesions (precisely placed damage) of particular parts of the brain. He correctly concluded three things.

Intellect is largely concentrated in the cerebral cortex.
The cerebellum is important for co-ordination of motor movements.
The lower brain sustains vital bodily functions.

However, he also insisted that mental functions cannot be dissociated from one another, and that removing cortex from an animal lowered its intellect in proportion to the amount removed.

Like other 19th century explorers mapping deeper into the "interiors", neuroanatomists also set out to localize the areas of brain functions. In the 1860s, **Gustav Fritsch** (1838-1927) and **Edouard Hitzig** (1838-1907) seemed to provide conclusive evidence of localization of cortical functions.

ELECTRICALLY STIMULATING PARTICULAR AREAS OF CORTICAL TISSUE CAUSES MOVEMENTS OF INDIVIDUAL LIMB OR FACE PARTS ON THE OPPOSITE SIDE OF THE BODY. *
AND LESIONING THE TISSUE RESULTS IN CORRESPONDING DISTURBANCES OF MOTOR FUNCTION.

* It has been known since antiquity that convulsions or paralysis following injury to one side of the head are manifested on the other side of the body.

Further support for cortical localization came in 1861. **Paul Broca** (1824-80) demonstrated that disorders of speech are associated with damage in a region of the left frontal lobe.

THE PERSON UNDERSTANDS WHAT IS SAID TO HIM BUT SPEAKS WITH DIFFICULTY, IF AT ALL.

This is known as **Broca's aphasia**. Broca's area co-ordinates the movements of speech. It is right next to the part of the **motor cortex** which controls movements of the lips, tongue and vocal cords.

In 1874, **Carl Wernicke** (1848-1904) discovered that damage to an area of the temporal lobe, close to tissue involved in hearing (the **auditory cortex**), resulted in another type of language disorder.

THESE PEOPLE SPEAK FLUENTLY BUT WHAT THEY SAY IS MOSTLY MEANINGLESS.

This is **Wernicke's aphasia**.

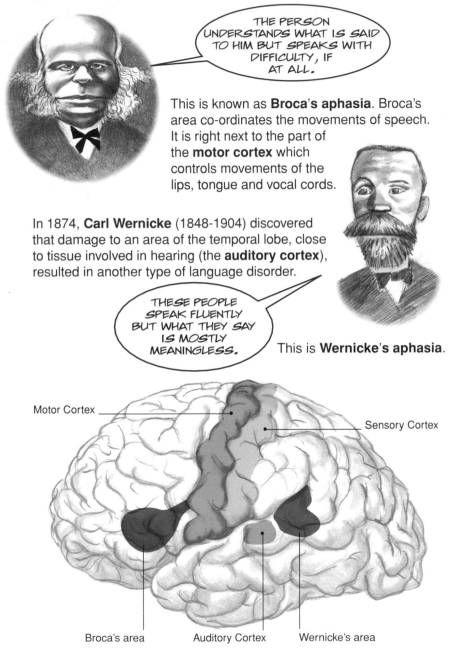

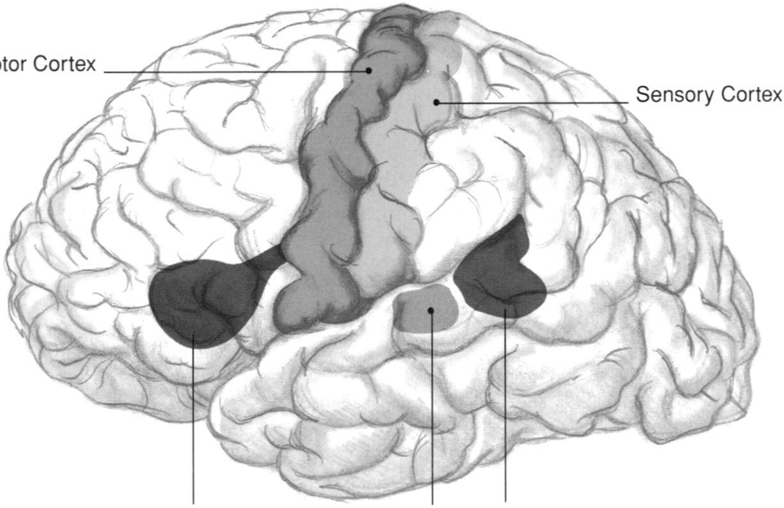

Many years later, the neurosurgeon **Wilder Penfield** (1891-1976) was able to use stimulation of conscious patients undergoing brain surgery* to map the human motor strip (or motor cortex) in the frontal lobe. He also charted the sensory strip in the parietal lobe.

* Remember, Aristotle already knew that touching the brain does not cause pain or any other sensation.

Despite these successes, the localizing of high level mental functions to particular patches of cortex faced continuing opposition. Not least because the localizers began to produce brain maps just as inconsistent with one another as those of the phrenologists.

I HAVE REMOVED THE ENTIRE CORTEX OF THIS DOG, YET HE STILL STANDS AND WALKS.
THEREFORE FRITSCH AND HITZIG MUST BE WRONG TO SAY THAT THE CENTRES FOR MOTOR MOVEMENT MUST BE IN THE CORTEX.
FOOD

Friedrich Goltz (1834-1902)

In the 20th century, **Goldstein** and **Lashley** followed **Flourens**'s and **Goltz**'s wholist view that higher functions depend on the **whole** cortex, and loss of function on the extent of tissue damaged. Others, such as **Monakow** and **Sherrington**, eventually abandoned materialism and identified higher mental functions with a soul.

Beginning to Assemble Brain Functions

One of the first people to see a solution to the apparent contradiction between localization and wholism was **John Hughlings-Jackson** (1835-1911). Hughlings-Jackson accepted that simple sensory and motor functions are localized within specialized cortical areas. But he also saw that more complex thinking and behaviour must be **assembled** from a great many of these simpler components, so involving many separate areas of the brain. He also understood that the "same" activity can be assembled at **lower or higher levels** of the brain.

An infant held up by the hands exhibits walking controlled by the spine. Yet when it is older, it will still have to learn voluntary "cortical" walking.

AND THOUGH BROCA'S PATIENTS CANNOT TALK, THEY CAN SOMETIMES CURSE OR SING. THESE AUTOMATIC REACTIONS TO A STUBBED TOE OR THE SOUND OF A TUNE MUST ORIGINATE FROM SUB-CORTICAL CENTRES.
THEY DO NOT REQUIRE THE CORTICAL REGIONS WHICH ARE NECESSARY FOR THE CONSTRUCTION OF VOLUNTARY, NON-AUTOMATIC SPEECH.

Hughlings-Jackson and, later, **Henry Head** (1861-1940) recognized that just because our vocabulary contains single words such as "walking", "speaking", "seeing" or "remembering" does not mean that these refer to single activities.

The great Russian neuropsychologist **Alexander Luria** (1902-77) pointed out that the same function can be achieved by different **ensembles** of brain areas working together on different occasions. For example, learning a new skill requires conscious, cortical thought. But control may subsequently pass down to sub-cortical centres once the skill is well learnt.

INDEED, CONSCIOUSLY THINKING ABOUT A WELL-LEARNT SKILL MAY ACTUALLY DISRUPT IT.
LEFT?
BUT THE TRAFFIC LIG
IS RED!
WHERE ARE THE
BRAKES?
A CHILD IS... OH, GOD!
THAT TRUCK
TOO CLOSE !!!

NOW... TURN LEFT.

Tracing the Progress

Is the brain made up of blood vessels, glands or globules? This debate in the 17th century could only progress with better techniques for visualizing a **dense** and **three-dimensionally** complex organ. Technical advances included: improving neuroanatomy; tools for dissection; developing chemical means for fixing and preserving brain tissue; refinement of microscopy; inventing tissue-staining techniques.

The cell theory of the nervous system was established by the end of the 19th century.

Neurons and Glia

In fact, there are two types of cell in the brain: **neuron(e)s**, of which there are 100,000,000,000; and even more **glial** cells. Neurons, or nerve cells, are what is normally meant by "brain cells". There are many types of neuron. All of them have a **cell body**, an **axon**, and lots of branching fibres called **dendrites**.

Rather little is known about glial cells. One thing they do is to produce **myelin**, a fatty insulating substance that sheathes many axons. Depletion of myelin is a feature of several neuro-degenerative diseases, such as Multiple Sclerosis.

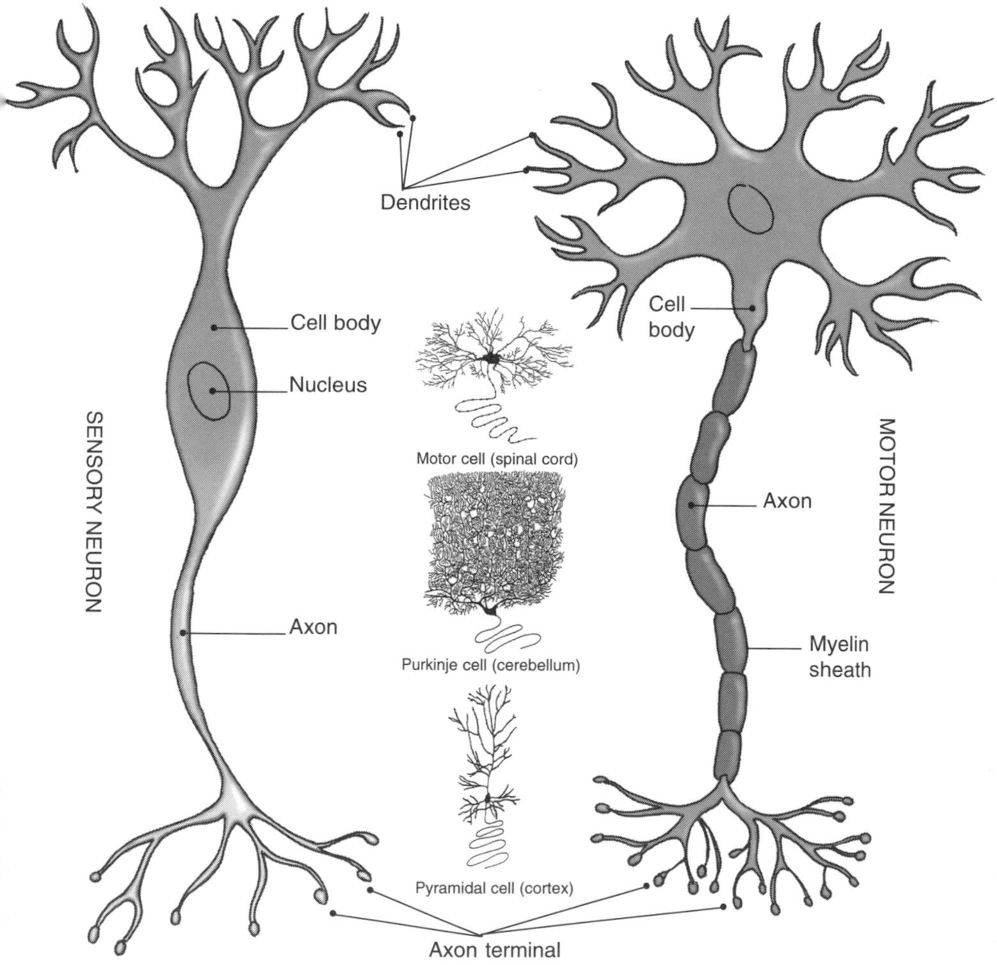

Grey and White Matters

Where a lot of cell bodies are packed closely together they appear as "**grey matter**", or **cortex**. Where the tissue is mainly long myelinated axons connecting different communities of cells (known as **nuclei**), it appears as "**white matter**".

The convolutions of the cortical surface cause much of it to be hidden inside folds known as **sulci** (**sulcus** in the singular), which are separated by ridges known as **gyri** (**gyrus**).

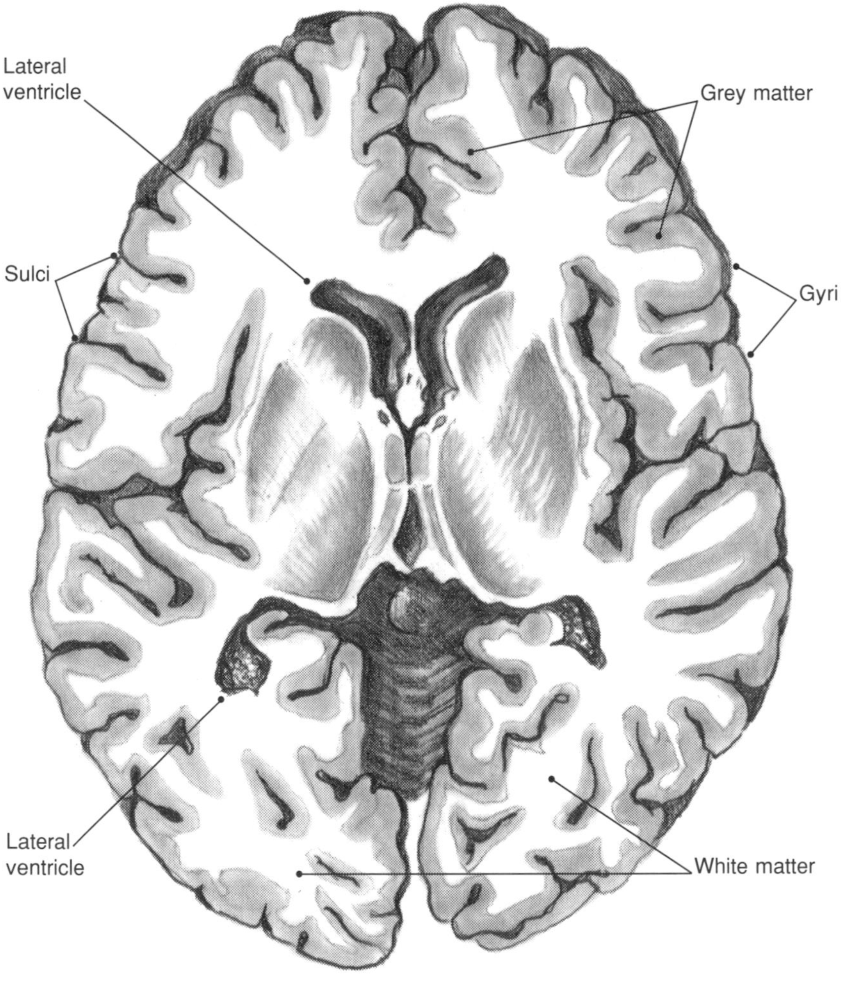

The Electric Brain

Neurons have the property of “nervous irritability”, meaning they respond to external stimuli, such as an electric current. If a cell body gets the right sort of stimulation/information from its dendrites and from the axons of other cells, it will “fire” (show irritability). This means it sends a small **electrical signal** down its axon. The axon then connects to the dendrites or cell bodies of other neurons or to the cells of **muscles** or **glands**.

Neuroscientists can study a neuron by placing **electrodes** close to the cell body.

A **recording electrode** monitors the number of times the cell fires each second. A **stimulating electrode** will drive the firing of the cell.

Every neuron is stimulated by huge numbers of other nerve cells connecting to its dendrites or cell body. Some of these connections are **excitatory** (they increase the likelihood of the target cell firing). Some are **inhibitory** (decreasing the likelihood of firing). The relative amounts of excitation and inhibition impinging on the target cell jointly determine its firing rate.

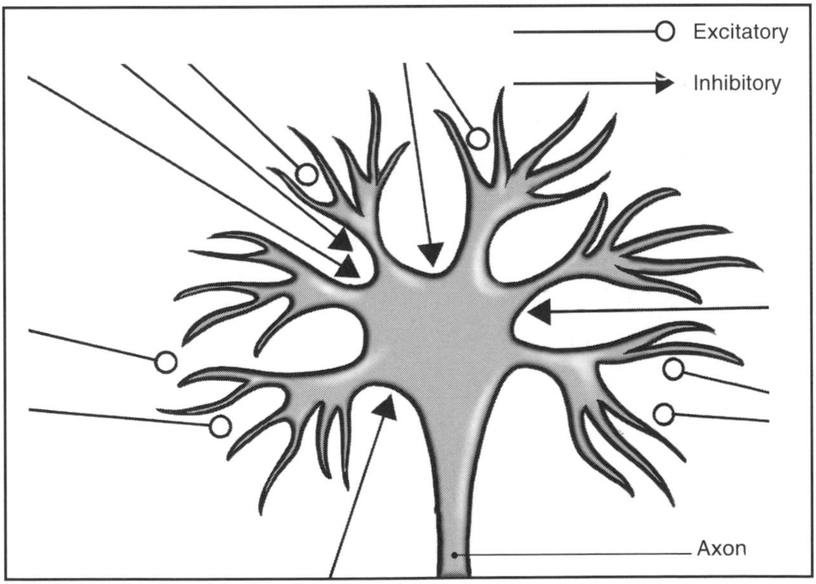

The figure shows a cell receiving excitatory connections (mainly to its dendrites) and inhibitory connections (mainly to its cell body)

Abnormal Firing

Sometimes the firing of groups of cells can become excessive.

THIS MAY BE EXPERIENCED AS MUSCLE TICS.
OR AS VISUAL DISTURBANCES, SUCH AS THOSE ASSOCIATED WITH MIGRAINE.
IN EPILEPSY, THE ONSET OF ABNORMAL FIRING MAY PRODUCE THE EXPERIENCE OF AN AURA.
BUT AS THE ABNORMAL FIRING EXTENDS TO MORE AND MORE TISSUE, IT RESULTS ULTIMATELY IN A SEIZURE.

The Chemical Brain

Where axon branches connect to the dendrites or bodies of target cells there is a small gap, which **Sir Charles Scott Sherrington** (1857-1952) named the **synapse**. The electrical potential coming down the axon cannot jump this gap. Instead, the **pre-synaptic** axon releases specially shaped chemical **molecules**.

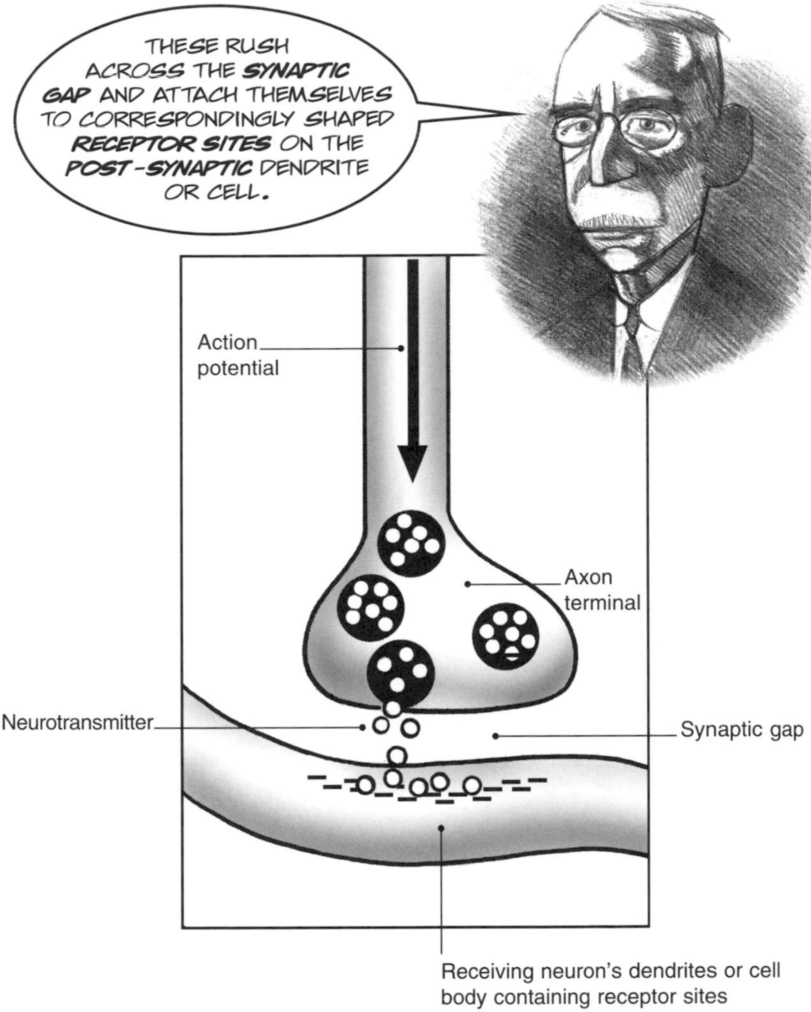

If the next cell is another neuron, arrival of the molecules will either increase (excitatory) or decrease (inhibitory) the likelihood of that cell firing.

Chemical Malfunctions

The chemicals which act in this way are known as **neurotransmitters**. Examples are **serotonin** and **dopamine**. Too little or too much of a neurotransmitter leads to malfunctions of various kinds. In Parkinson's disease, for example, voluntary movements become difficult to initiate and control. This is associated with a shortage of brain dopamine. Increasing production of dopamine in the brain improves the condition.

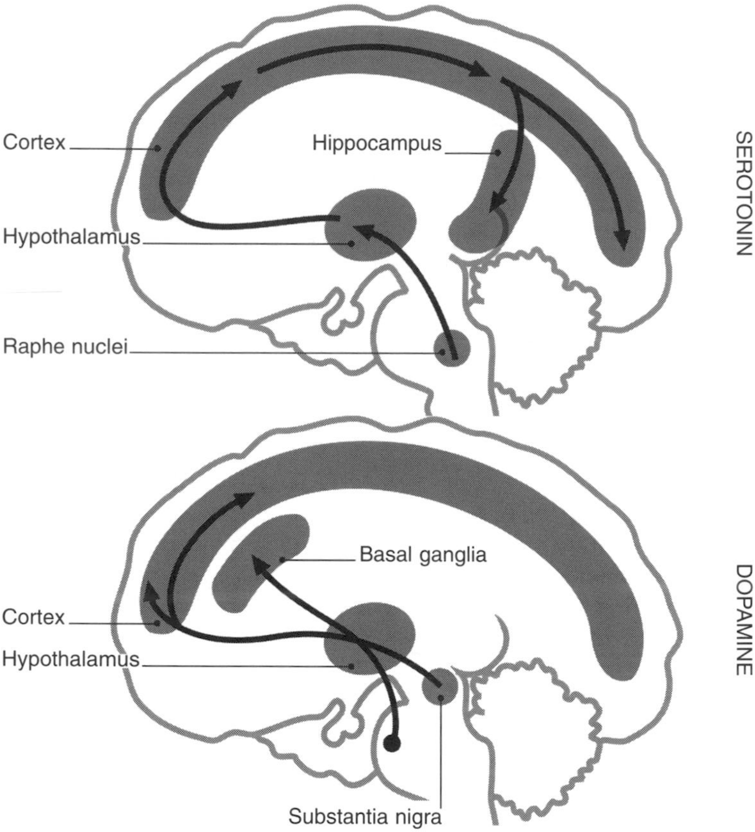

Why do drugs such as morphine and LSD, and poisons such as curare, have their effect? Because they have a similar structure to the brain's natural neurotransmitters. By attaching themselves to post-synaptic receptor sites, they disrupt normal traffic along the neural pathways.

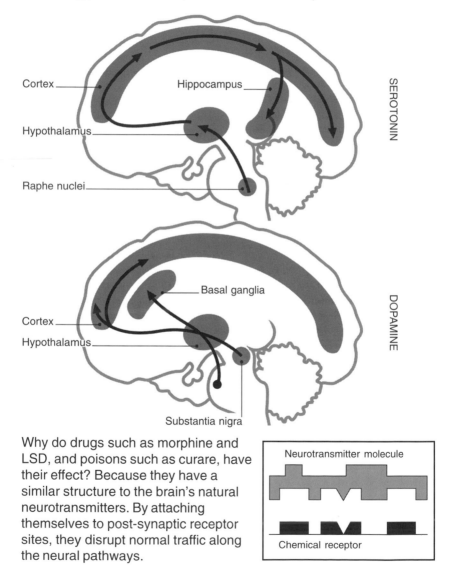

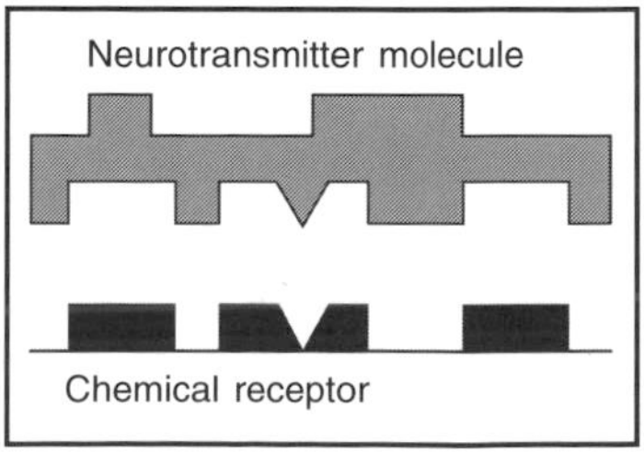

Brain, Hormones and Body

Neurotransmitters are similar in many respects to **hormones**. Hormones, such as **adrenaline** and **testosterone**, are secreted into the blood by glands. In the blood, they can travel to affect distant organs.

Hormones regulate bodily functions, such as energy production and metabolism.

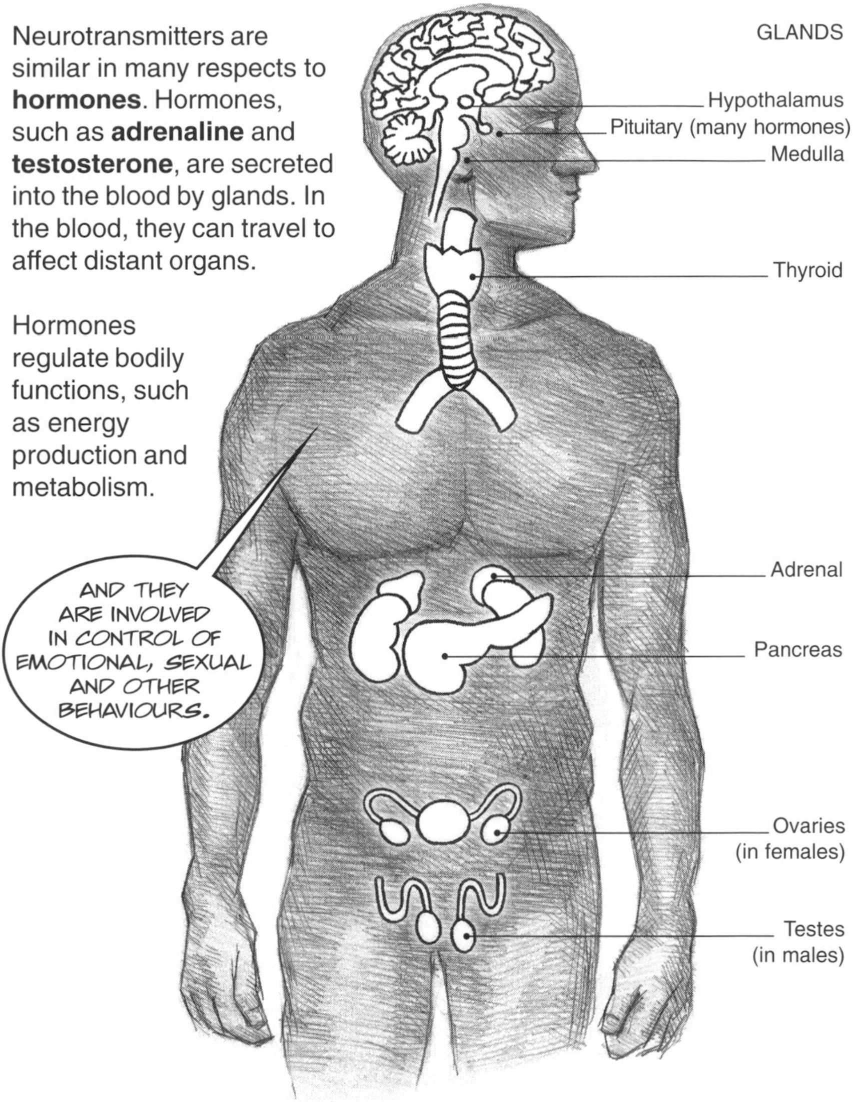

• Brain activity controls the release of hormones by the glands into the blood.
• But hormones, carried up to the brain in the blood, then serve to influence the activity of the brain itself.
• The brain is a bodily organ, part of a larger functioning system. When, as in this book, we focus exclusively on the brain, we easily lose sight of this fact.

The Geography of the Human Brain

The brain is a fantastically complex structure. The terminology that has evolved to describe it is almost more daunting still. Because the brain is studied by many different groups – anatomists, physiologists, biochemists, geneticists, surgeons, neurologists, neuropsychologists and others – most structures have acquired several alternative labels, which may be in Greek, Latin, English or French.

The naming of disorders of behaviour resulting from brain injury also raises difficulties. Many disorders have names starting with “a” meaning “without” (as in a-theism). Others begin with “dys” meaning “bad” (as in dys-lexia). Many of the “a”s should really be “dys” because it is relatively rare for a behavioural function to be completely eradicated. Although this happens sometimes, degrees of impairment are more common.

You have been warned!

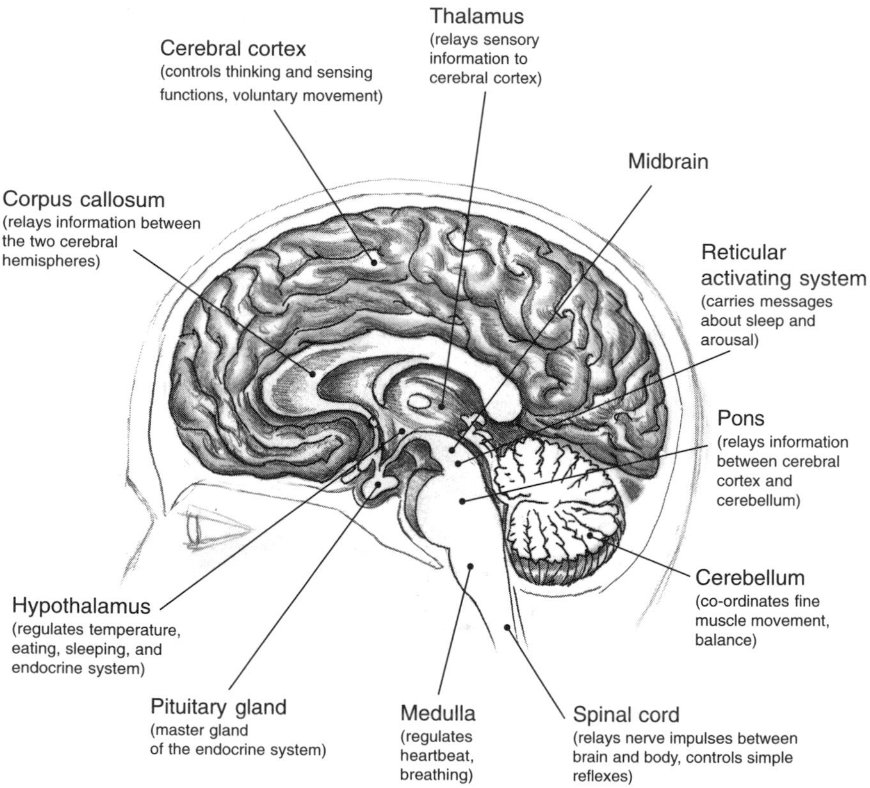

Evolution and Development

Nervous systems evolved because they improved the survival chances of animals that had them. A nervous system allows an animal to **behave** rather than be **passive**: to seek food and avoid danger rather than simply to hope that food will come along but that danger will not.

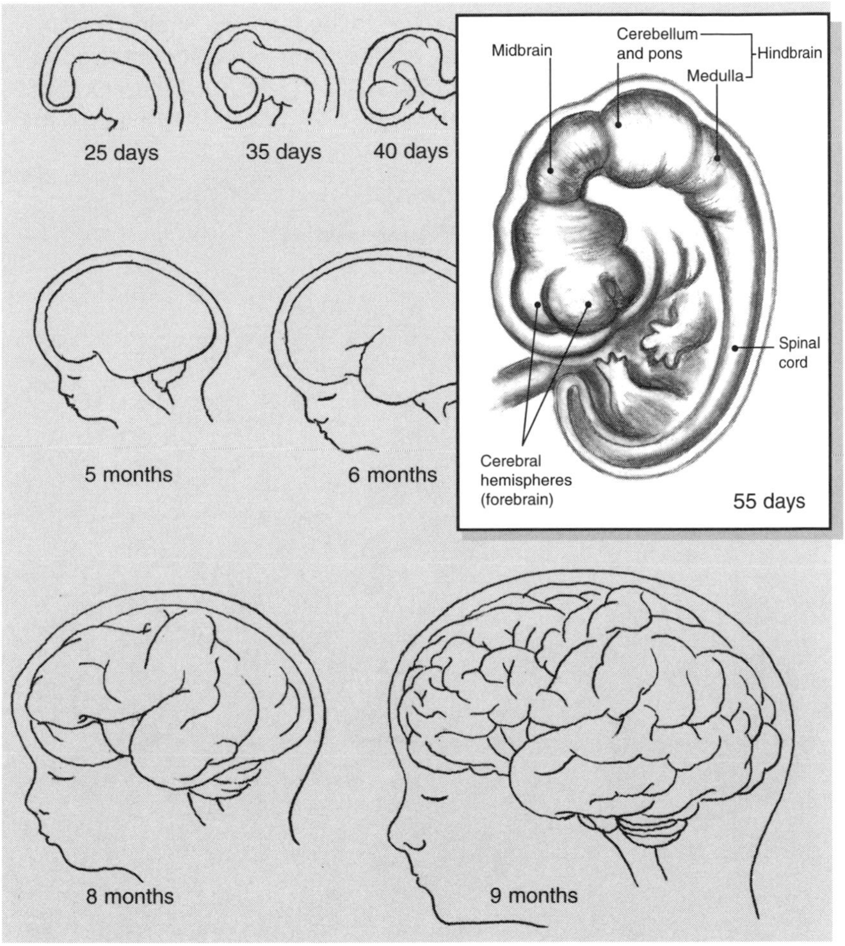

The brain of an embryo starts off as a simple tube of tissue. It then develops three enlargements that will become the **forebrain**, **midbrain** and **hindbrain**. The cortex of the forebrain later divides into the two **cerebral hemispheres**, which grow outwards to cover much of the lower brain regions.

The Hindbrain

The "lower" brain, or hindbrain, mainly supports vital bodily functions.

The first major component of the hindbrain is the **medulla**. It is a continuation of the spine and is concerned with control of breathing, heartbeat and digestion. Above it is the **pons** which receives information sent from visual areas to control eye and body movements. It sends this information to the third major structure of the hindbrain, the walnut-shaped **cerebellum**, which controls co-ordination of movement sequences. A fourth hindbrain structure, the **reticular formation**, is important in the control of arousal and in the sleeping and waking cycle.

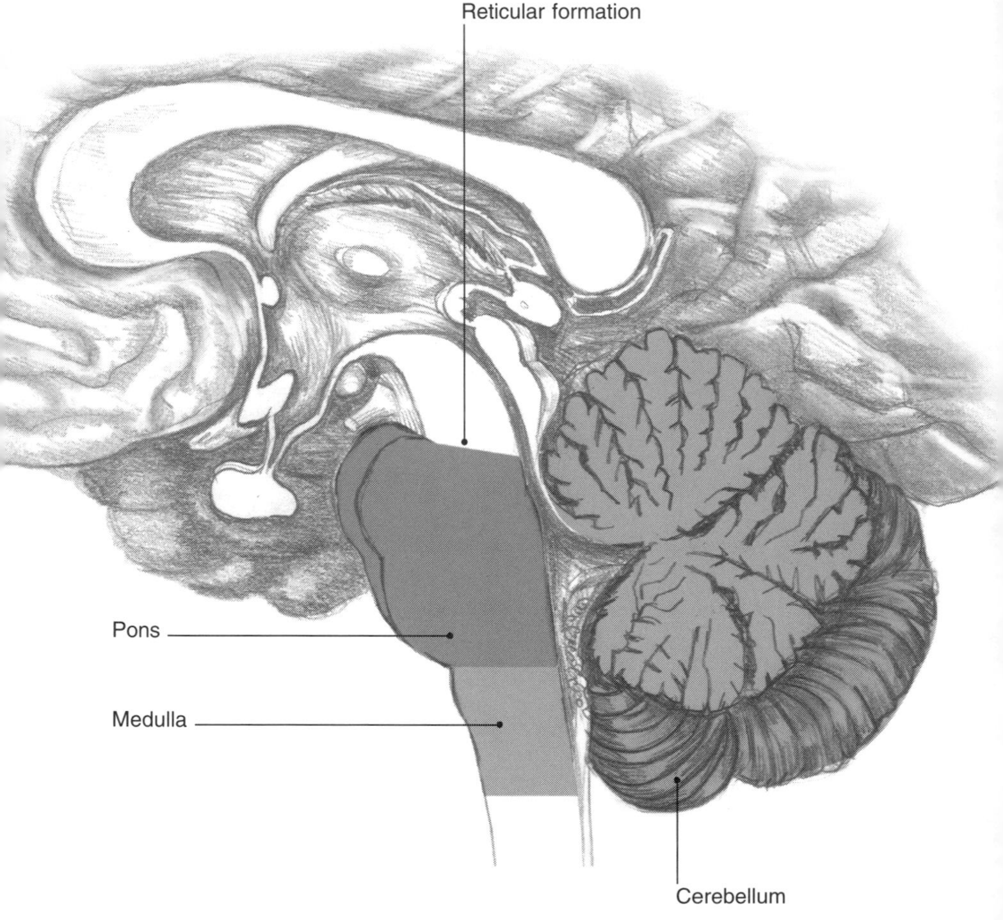

The Midbrain

The midbrain sits above the hindbrain. Its main components are the **basis pedunculi**, **tegmentum** and **tectum**. The first two of these are concerned with motor movements. It is dopamine shortage in the pedunculi, and elsewhere, which gives rise to Parkinsonism. The tectum contains visual and auditory **nuclei** (groups of cells). For birds and other lower animals, these **are** their visual and auditory brains. Mammals have evolved large areas of forebrain dedicated to these senses, but their tectums still govern whole body movements in response to light and sound.

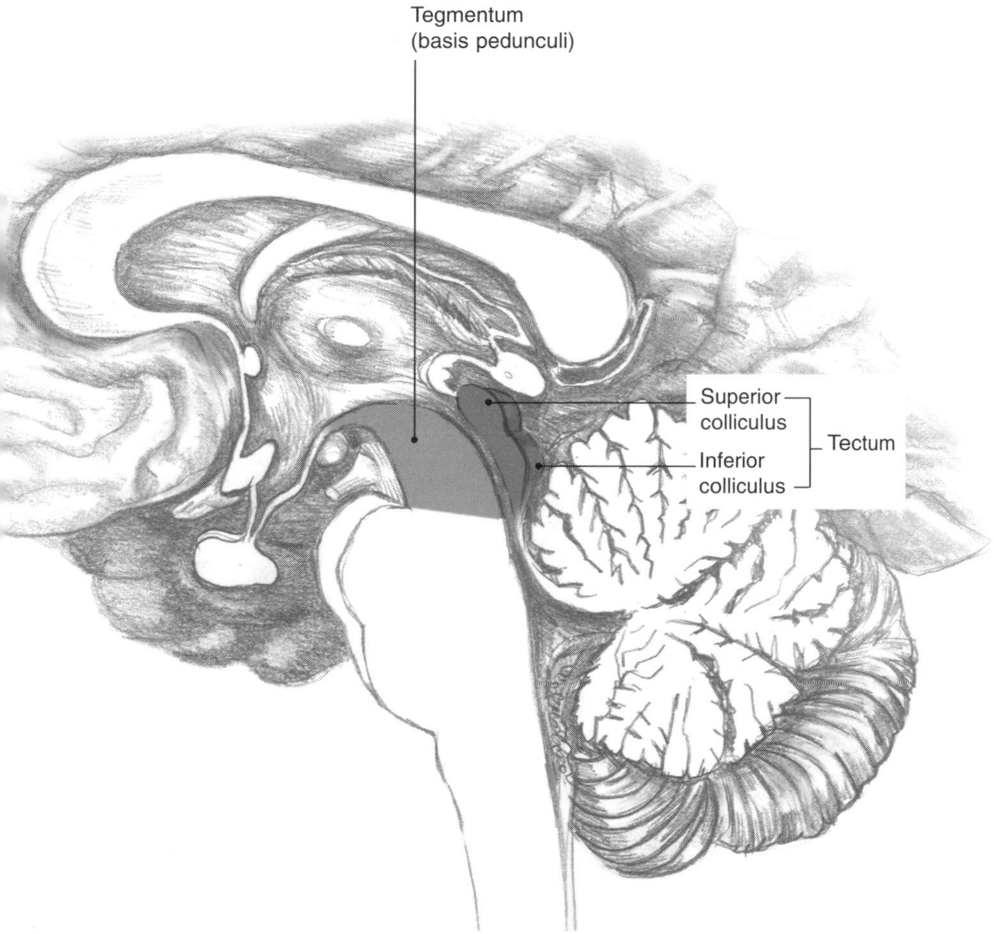

The Forebrain

The human forebrain contains a large number of important structures. The **thalamus** is a kind of communications centre which receives input from eyes, ears, skin and other sensory organs. It also modulates activity in the cortex as a whole. The **hypothalamus** is a small but very complex structure engaged in control of the four Fs (feeding, fighting, fleeing and fornication), as well as of temperature regulation, sleep and the expression of emotion.

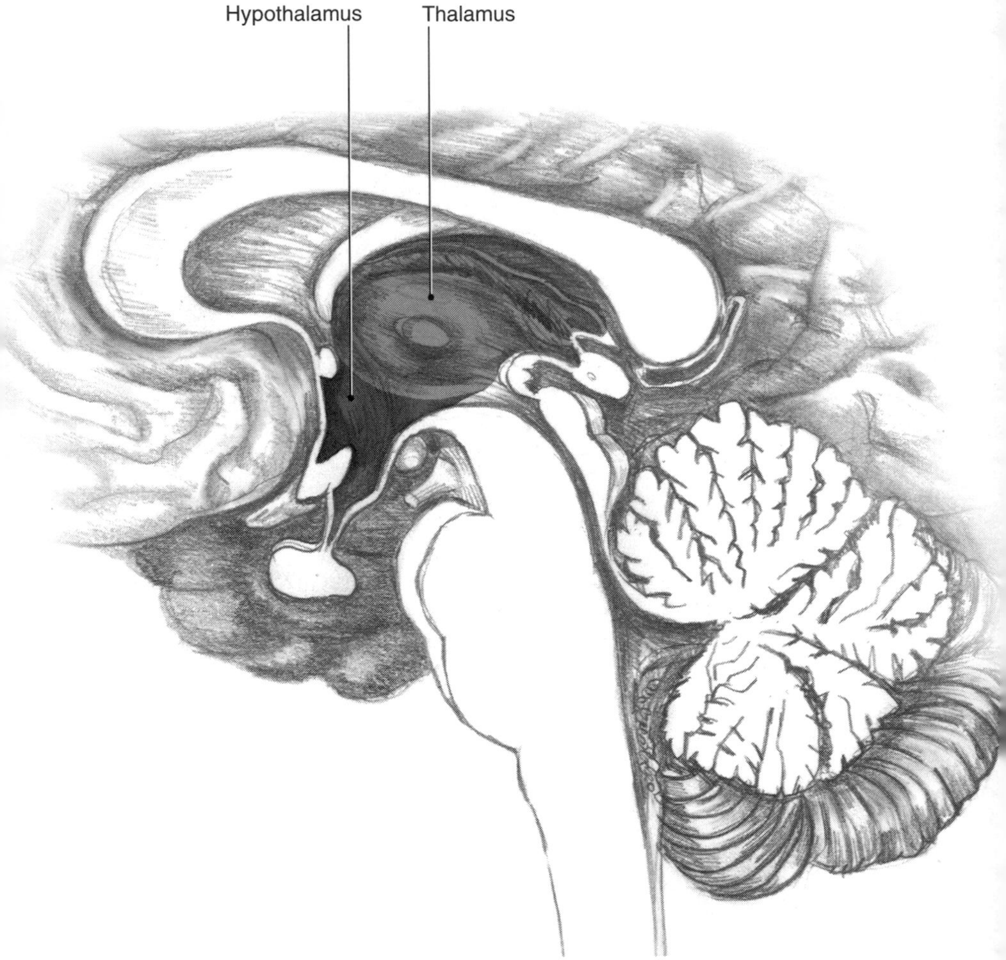

The **limbic system** began as a "smell brain" and is heavily involved in emotional processes. The **hippocampus** in the limbic system is essential for knowledge about the spatial layout of the environment.

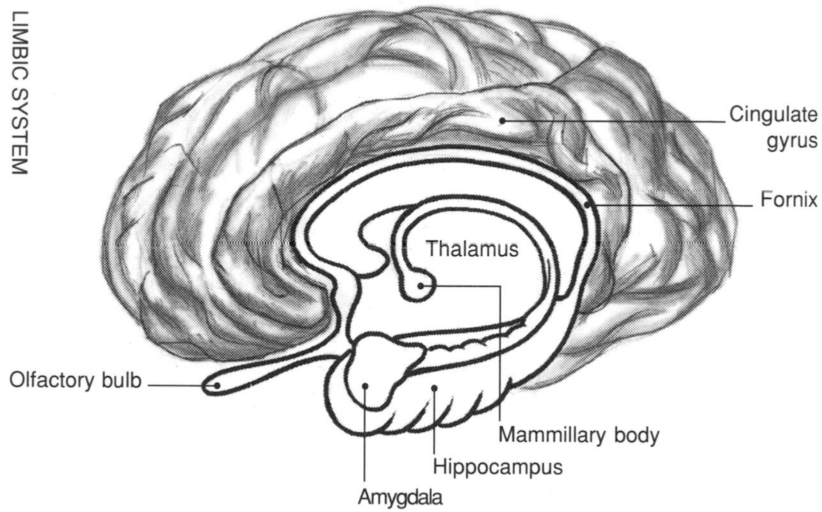

The **basal ganglia** are a number of nuclei (grey matter) which play a major role in movements. People with Parkinsonism show dopamine shortages here too. Distinctive regions of the basal ganglia receive inputs from either the limbic system or from various cortical areas. It may be here that emotions and memories compete with present circumstances and thoughts for control of behaviour.

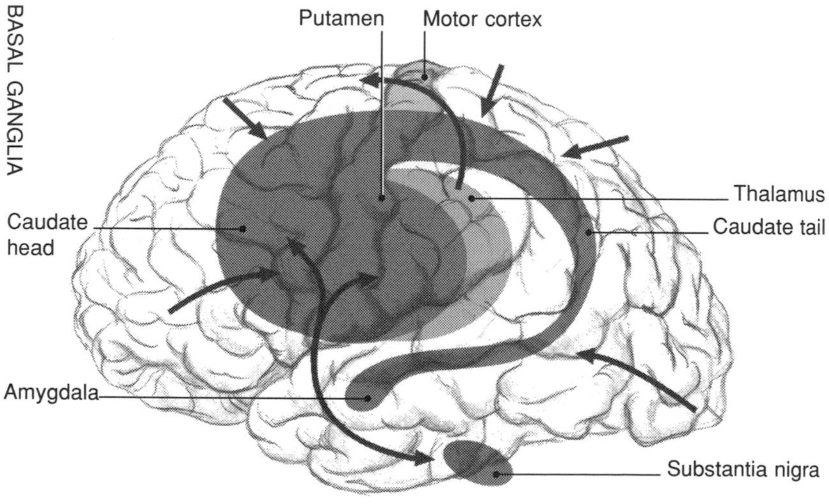

Left and Right Hemispheres (LH and RH)

The two **cerebral hemispheres** are the largest and most obvious features of human and other primate brains. Their surface grey matter is **the cortex,** sometimes called **neo-cortex** to distinguish it from the **cortex** found in lower, and more ancient, brain structures. Each hemisphere receives information largely from the opposite side of the body, which it also largely controls. The two hemispheres can act in concert to produce coherent behaviour because they share information via a vast sheet of fibres known as the **corpus callosum**. They are also indirectly connected through the sub-cortical structures on top of which they sit.

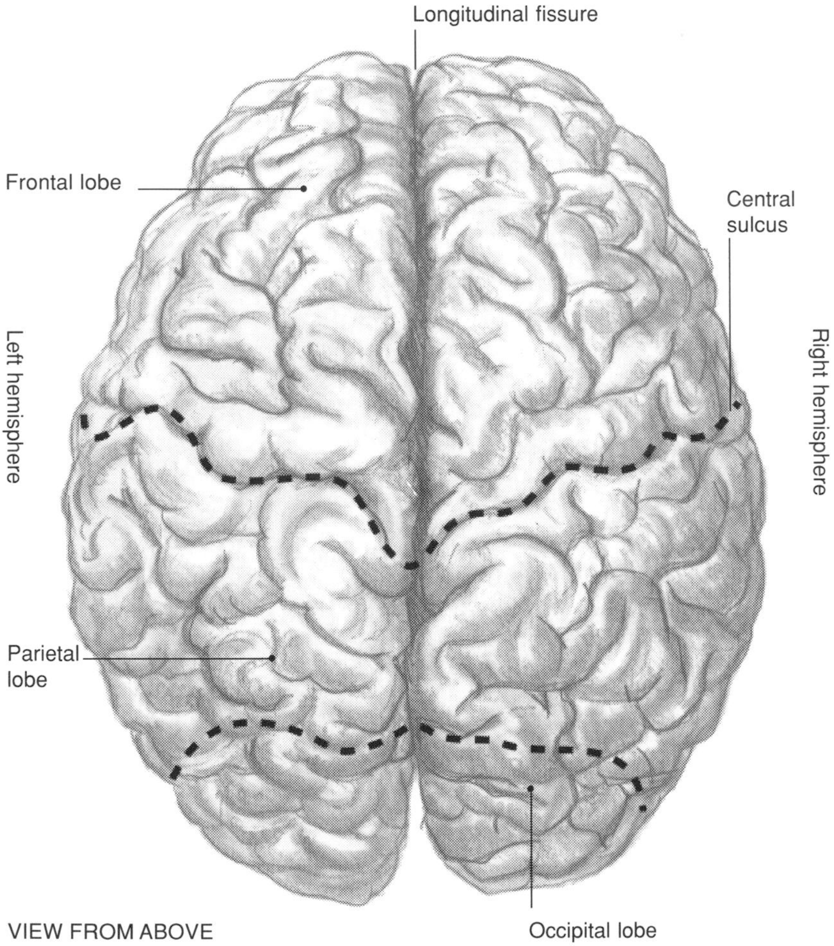

Each hemisphere is divided into four **lobes**, separated by deep clefts known as **fissures**. The lobes in turn can be divided into areas. Different areas are identified on the basis of a number of criteria. They look different when stained and seen under the microscope and are distinguished by the pattern of their connections to other areas. They are **functionally defined** by the type of stimuli which activate their cells and by the impairments in behaviour which result when they are damaged.

The identification of areas remains a live topic of investigation. Identifying **corresponding areas** in the brains of **different species** is particularly difficult.

LATERAL VIEW

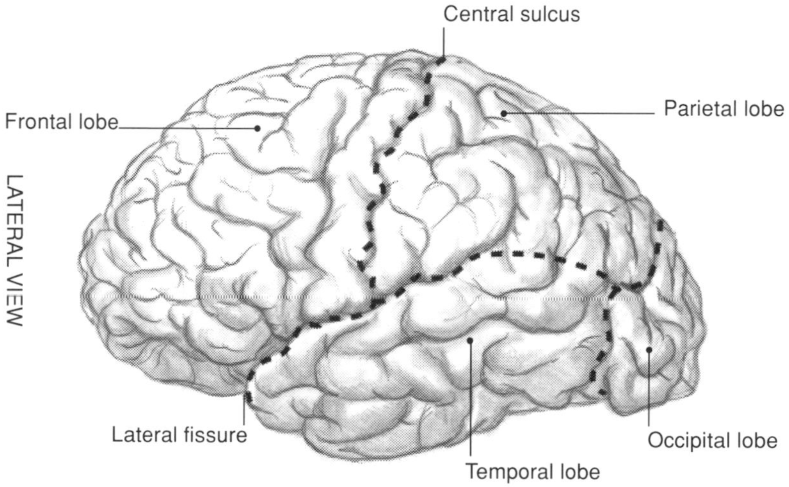

MEDIAL VIEW

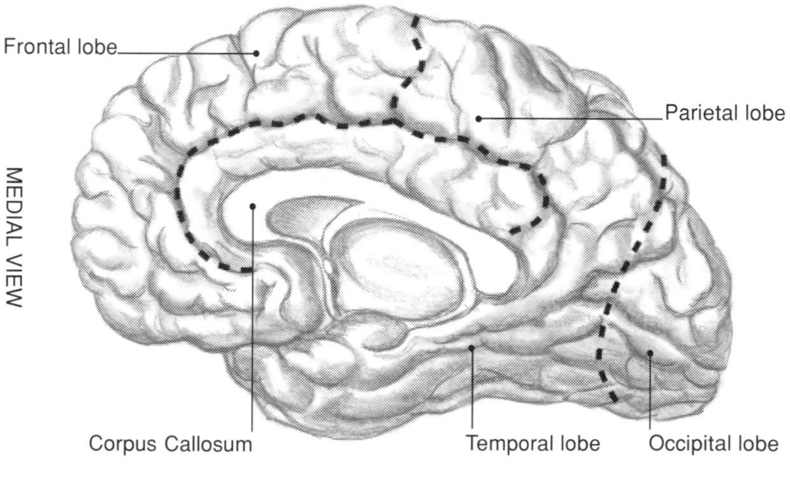

Mental Abilities

The cerebral **cortices** (plural of cortex) are the site of the most **advanced mental abilities**.

They contain centres that put together information from the **senses** with **thoughts** and **memories** to work out what is going on in the world about us.

Primates, and humans in particular, have hemispheres that are specially massive.

It is important to bear in mind, however, that the cerebral cortices function as part of a larger system. **Connectivity** is an all-important feature of the brain. Higher and lower centres are strongly connected by ascending and descending fibre tracts. These maintain contact between the structures of the hindbrain, midbrain and forebrain. **This is how integration of mind and body is achieved**.

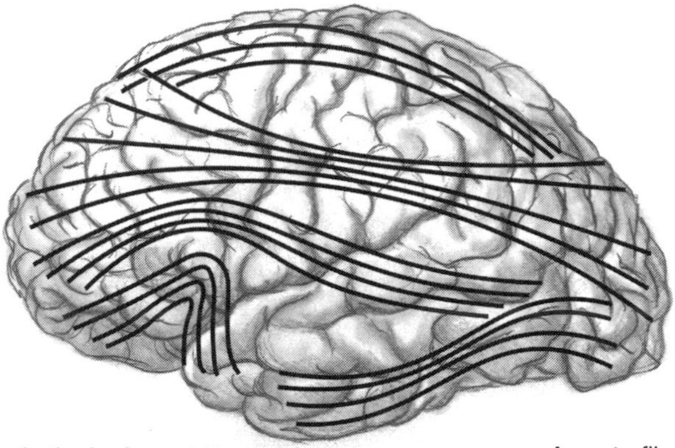

Connectivity in the brain

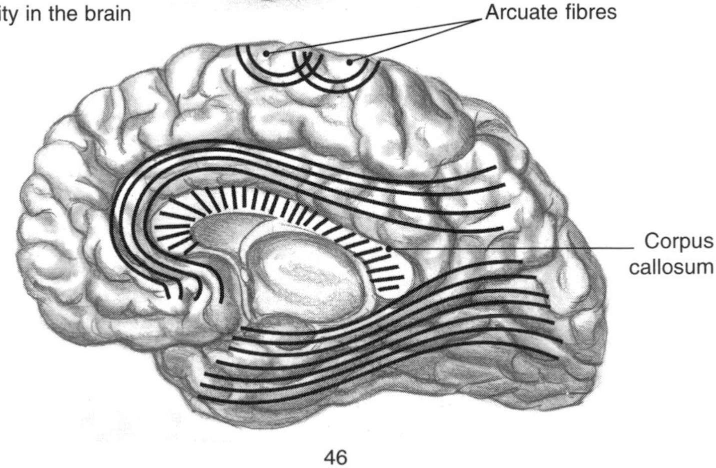

Simple Minds 1: the Sea Slug

Some behaviour appears more complicated and intelligent than it really is.

If you try reading near a loud clock, its ticking may distract you, making it hard to concentrate on your book.

This learning to ignore a stimulus is known as **habituation**. The humble sea slug, **Aplysia**, is capable of habituation. When its head is touched with a glass rod, it responds with a defensive withdrawal of its gill. But if this sequence is repeated often enough, the gill withdrawal response habituates.

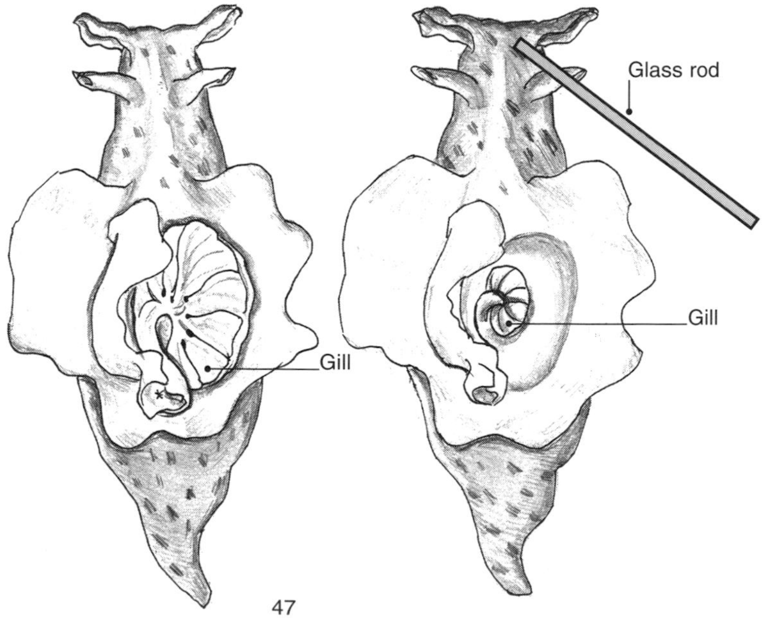

Return in imagination to the room with the loud clock which you have learnt to ignore. Suppose you are now told there is a time-bomb in the vicinity.

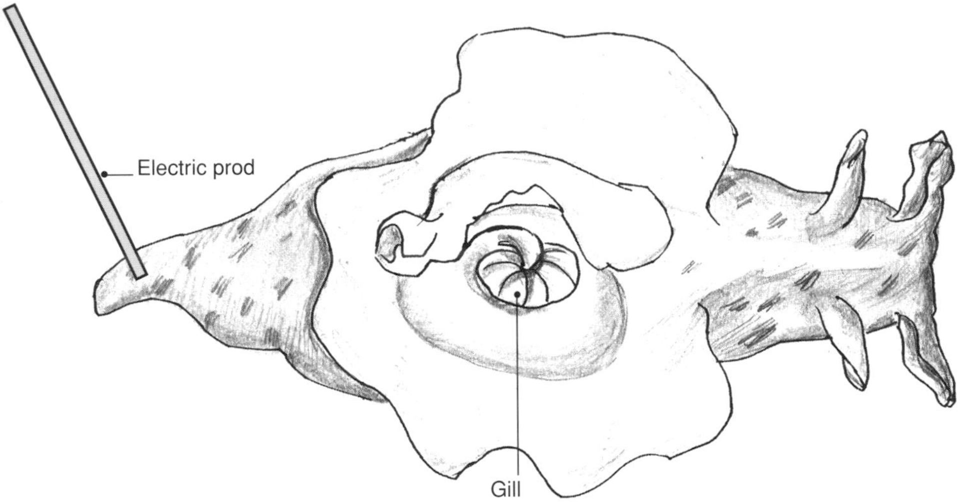

If Aplysia's gill withdrawal is habituated and the animal is then given a mild shock to the tail, the gill withdrawal response returns at great strength. Aplysia, too, shows sensitization.

Habituation and sensitization in the human example provoke use of mentalistic terms like **learning**, **attention** and **memory**. Yet we find similar behaviours in Aplysia which possesses a mere 5,000 neurons.

Simple Minds 2: Frogs and Toads

The eye of the **frog** contains cells that only fire in response to small, erratically moving dark spots. It is surely no coincidence that frogs try to eat a moving fly, but starve if surrounded by motionless dead flies.

ONLY A ***STIMULUS*** CLOSELY RESEMBLING A FLY ON THE WING WILL ***TRIGGER*** MY HUNTING BEHAVIOUR.

Toads will attempt to eat a match moving lengthways but show no interest in a match moving on its end.

AS FAR AS I'M CONCERNED, ANYTHING THIN, ELONGATED AND MOVING LENGTHWAYS QUALIFIES AS A ***WORM***, REGARDLESS OF COLOUR, TEXTURE OR RIGIDITY.

Simple Minds 3: Birds

When an adult **herring gull** has a grub in its beak, young chicks respond by gaping their mouths wide and cheeping excitedly. This might seem like intelligent behaviour on the part of a hungry youngster seeing food. But herring gull chicks are not very intelligent.

WE SIMPLY RESPOND TO THE RED SPOT ON THE ADULT'S YELLOW BEAK.

Paint the red spot yellow and the chicks ignore the food. Show them an empty beak with a red spot and they gape and cheep as before. In fact, a bright red dot on a vivid yellow pencil elicits extra strong gaping and cheeping. It acts as a **super-stimulus**.

Adult birds are no smarter than their offspring. Returning to the nest from foraging, they push food in the direction of the **largest** and **reddest** mouth in the nest. The success of baby cuckoos is due to their having bigger mouths and more crimson throats than the young of the host species in whose nests they find themselves.

WE FAIL TO NOTICE THAT THE INTERLOPER IS THE WRONG SIZE AND COLOUR. WE FEED IT IN PREFERENCE TO OUR OWN YOUNG.

The **throat** of the cuckoo chick is a **super-stimulus** for the release of the adult's feeding behaviour.

Simple Minds 4: Human Beings

Point-light displays demonstrate that only a small fraction of the available information may determine human perception and behaviour. An actor, black-faced, dressed in black and with light-emitting diodes positioned only at each **joint** of the body and limbs, is video-taped in high contrast. When the tape is played back, only the lights are visible.

While the actor is stationary, a viewer sees only a random array of lights. But as soon as the actor moves, observers perceive **a species-specific human movement pattern**, walking, running, dancing or whatever.

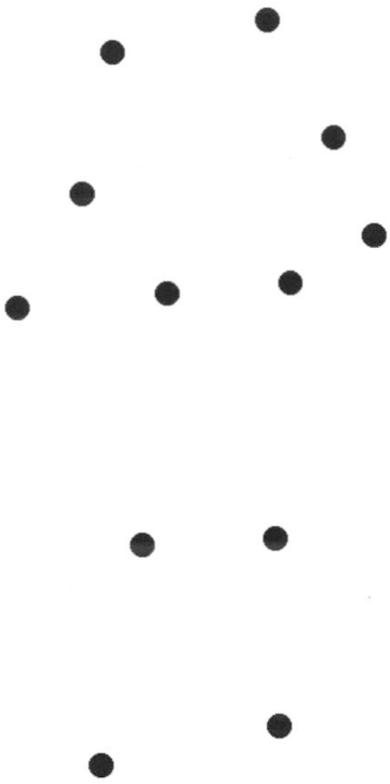

Stationary percept

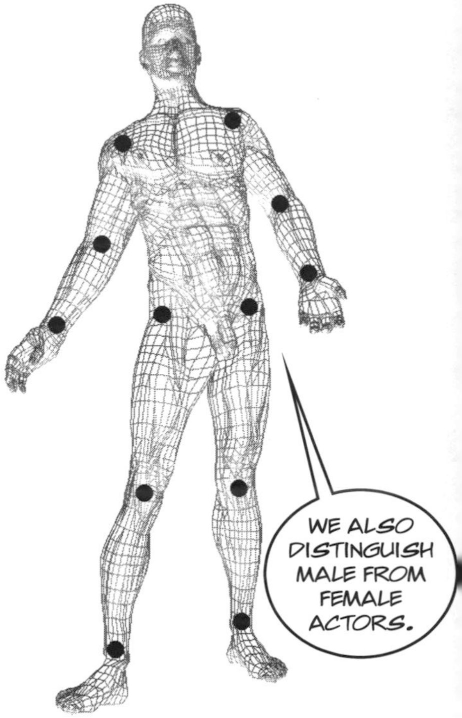
WE ALSO DISTINGUISH MALE FROM FEMALE ACTORS.

Kinetic percept

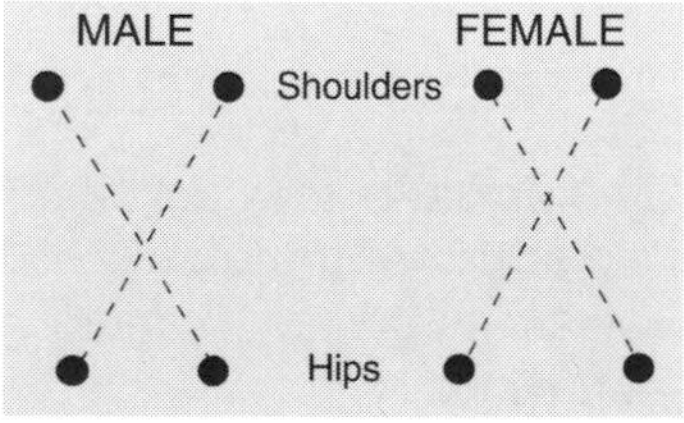
MALE
FEMALE
Shoulders
Hips

Selective removal of lights reveals that the discrimination depends on the **ratio** of **shoulder** swing to **hip** swing. The ratio is higher for males because their shoulders are broad relative to their hips.

These results show that our visual systems can recognize **species members** and their **genders** irrespective of facial features, hair or clothes, and with minimal information about body shape. Men swing their shoulders when trying to emphasize their masculinity, and women swing their hips when signalling their femininity. These are unconscious attempts to become a **super-stimulus** for **gender recognition**.

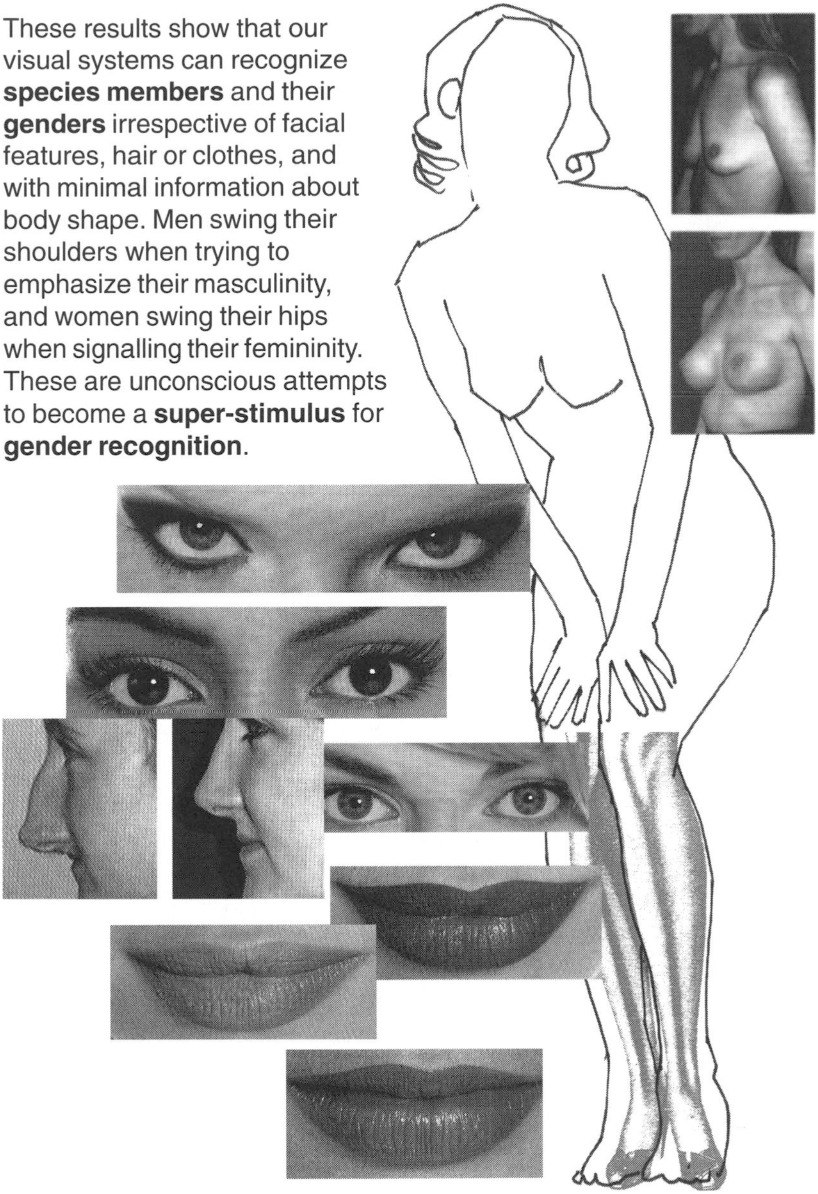

The cosmetic exaggeration of facial features – eyes, mouth, cheek-bones – for sexual effect is a very old and lucrative tradition. Padded and uplift bras, breast implants, bustles and cutaway swimsuits that illusorily enlarge the buttocks and lengthen the legs doubtless reflect cultural preferences. But they are all "glamour" exaggerations of nature and testify to human susceptibility to super-stimuli.

Complex Minds and Computers

Just as seemingly intelligent behaviours may turn out to rely on relatively simple mechanisms, so seemingly simple abilities can prove to be extremely complex.

In the early days of **computers**, people thought it would be easy to programme them to **recognize faces** and **words**.

1946 EDSAC computer
WE ALSO THOUGHT MACHINES MIGHT NEVER HAVE SUFFICIENT INTELLIGENCE TO TAKE ON THE INTELLECTUAL DEMANDS OF PLAYING CHESS OR PROVING MATHEMATICAL THEOREMS.
Gary Kasparov

Exactly the opposite turns out to be the case. Computers now beat the best human chess players and have devised novel mathematical proofs. When it comes to **walking** and **recognizing**, however, computers lag behind the young of almost any species you care to name. It has been chastening to discover that the problems proudly solved by human intelligence are simplicity itself in comparison to the problems solved by **evolution**.

Language and the Brain

Any attempt to understand the relationship between brain and mind has to face the issue of whether, or to what degree, *mental functions* can be localized to *particular* brain areas. Language has figured prominently in this debate, and no attribute of mind exhibits more clearly both the power and the limitations of a localizing approach to brain function.

By the end of the 19th century, Broca and Wernicke had established a special role for the left hemisphere (LH) in language (for right-handed people).

COOKIE JAR

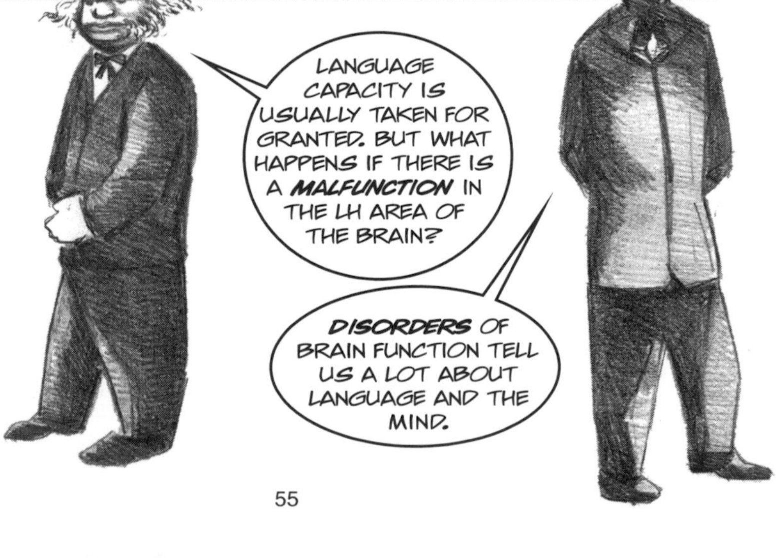
LANGUAGE CAPACITY IS USUALLY TAKEN FOR GRANTED. BUT WHAT HAPPENS IF THERE IS A ***MALFUNCTION*** IN THE LH AREA OF THE BRAIN?
DISORDERS OF BRAIN FUNCTION TELL US A LOT ABOUT LANGUAGE AND THE MIND.

Disorders of Language: the Aphasias

The aphasias are disorders of speech production or understanding. Here we examine the attempts of three aphasics to describe a picture. Each one suffers from a different type of aphasia. The first is a **Broca's aphasic**.

COOKIE JAR
COOKIE JAR ... FALL OVER ... CHAIR ... WATER ... EMPTY.
THE SPEECH OF THIS APHASIC LACKS GRAMMATICAL STRUCTURE AND FUNCTIONAL WORDS SUCH AS "AND", "IN" OR "HERE".
IT CONSISTS ALMOST ENTIRELY OF CONCRETE NOUNS AND VERBS, ALTHOUGH FOR SOME PATIENTS EVEN VERBS MAY BE IN SHORT SUPPLY.

Contrary to the classical doctrine of Broca himself, the impairment tends to be relatively mild unless damage extends beyond "Broca's area", in the **neo-cortex**, to include **sub-cortical** structures which co-ordinate speech.

Speech requires exquisitely detailed sequences of movements which must meet the constraints of **grammar** and **phonology** (why "weight" can be a word in English but "thgiew" cannot).

It is no coincidence that Broca's aphasics have trouble with verbs rather than nouns. It makes sense that the means for **naming** actions – verbs – are stored in the same cortical neighbourhood as the means for **controlling** actions. Here we have an insight into an important component of mind – *movement* itself.

The second type is a **Wernicke's aphasic**.

COOKIE JAR
WELL THIS IS ... MOTHER IS AWAY HERE WORKING HER WORK OUT O' HERE TO GET HER BETTER, BUT WHEN SHE'S LOOKING, THE TWO BOYS LOOKING IN THE OTHER PART. ONE THEIR SMALL TILE INTO HER TIME HERE. SHE'S WORKING ANOTHER TIME BECAUSE SHE'S GETTING TO. SO TWO BOYS WORK TOGETHER AND ONE IS SNEAKING AROUND HERE, MAKING HIS WORK AN' HIS FURTHER FUNNAS HIS TIME HE HAD.
THESE APHASICS SPEAK FLUENTLY IN WELL-FORMED AND PROPERLY INTONED SENTENCES, BUT WHAT THEY SAY LACKS MEANING AND CONTAINS WRONG WORDS OR EVEN NONSENSE WORDS.

A Wernicke's aphasic has lost *comprehension*. She understands neither what she says nor what she hears. But just as normal sentence structure and intonation are maintained, so are other **linguistic conventions** such as body language and turn-taking in conversation.

COOKIE JAR
LL
E
T
H
INTERESTINGLY, USERS OF SIGN LANGUAGE WHO HAVE A STROKE IN WERNICKE'S AREA SHOW THE SAME PATTERN OF IMPAIRED SIGNING AND COMPREHENSION OF SIGNING.
DAMAGE IN THIS TYPE OF APHASIA CENTRES UPON THE TEMPORAL LOBE AREA WHICH I IDENTIFIED.

As with Broca's aphasia, however, the condition tends to be fairly mild unless damage extends into surrounding areas. In addition, occasional individuals who sound like Broca's or Wernicke's cases turn out to have their damage in totally the "wrong" area. It might be said that the two most famous aphasic disorders actually license only **two cheers for localization**.

The third type is an **Anomic aphasic**.

The anomic aphasic also produces reasonably grammatical sentences but, because he has difficulty in finding words, he hesitates and uses indefinite nouns such as "thing".

His problem is most acute when he has to name objects without any context of use or talk. When shown a **pen**, he might be unable to name it.

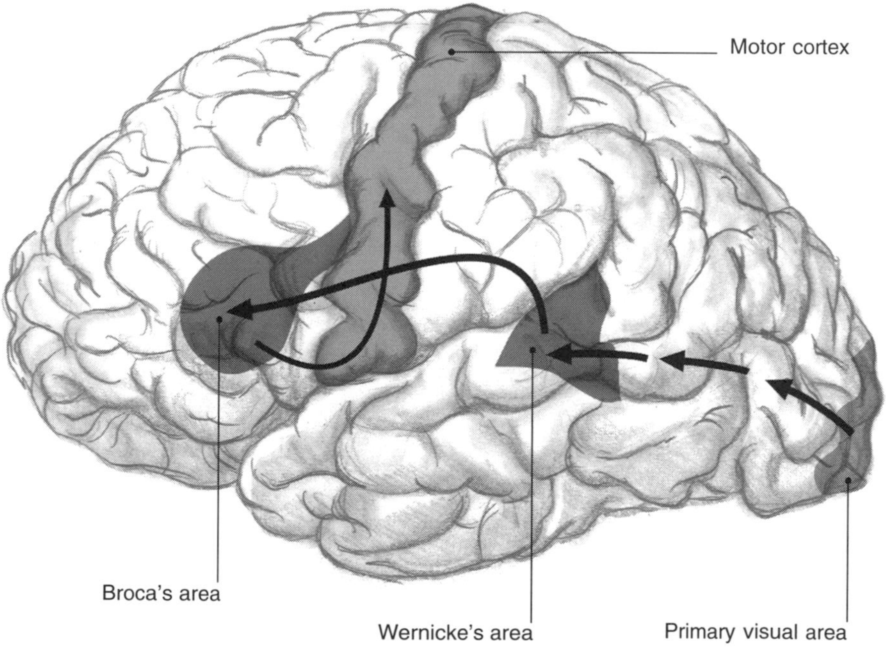

We saw that anomia for verbs tends to follow damage to frontal regions involved in the control of action. Correspondingly, anomia for nouns often results from damage to the temporal lobe which plays a major part in object recognition. The ability to **name** objects seems to be located close to the ability to **recognize** them. The logic of this arrangement goes even further.

Some anomics lose the names for particular categories, such as fruits, animals, or **colours**.

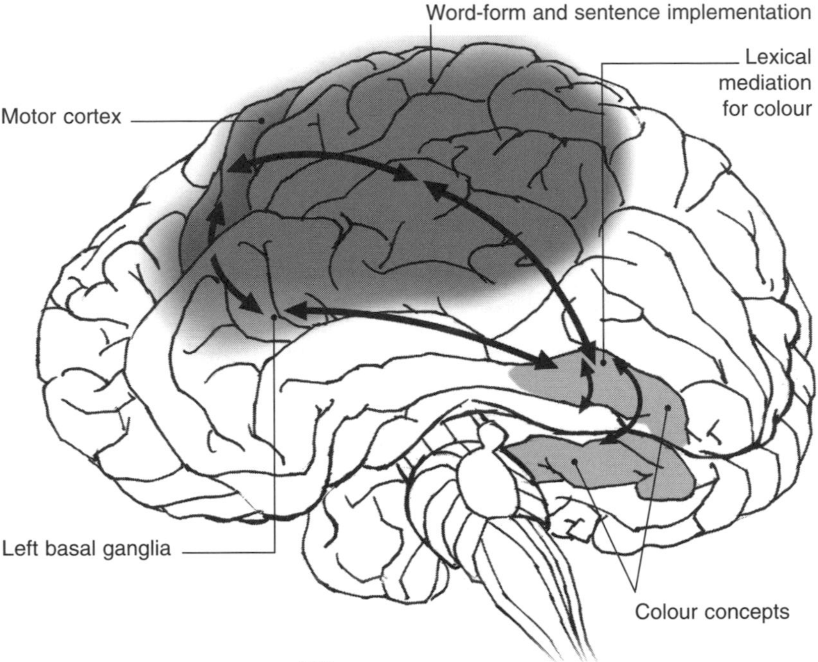
Word-form and sentence implementation
Lexical mediation for colour
Motor cortex
Left basal ganglia
Colour concepts

LOSS OF COLOUR NAMES OCCURS WITH DAMAGE TO THE VERY POSTERIOR (BACK OF THE) LEFT TEMPORAL LOBE.
THAT PUTS IT CLOSE TO THE REGION OF THE OCCIPITAL LOBE THAT SPECIALIZES IN PERCEPTION OF COLOURS.

A Model of Language Use

Wernicke suggested a model of language use that attempted to explain the aphasias and other language disorders. When we want to speak a thought, the words for it are put together in Wernicke's area and sent, via a bundle of fibres called the **arcuate fasciculus,** to Broca's area. Here the correct sequence of speech movements is called up and sent to the nearby motor cortex which carries them out. Wernicke's model is a sequence: **thoughts** into **words** into **sounds** into **muscle commands**.

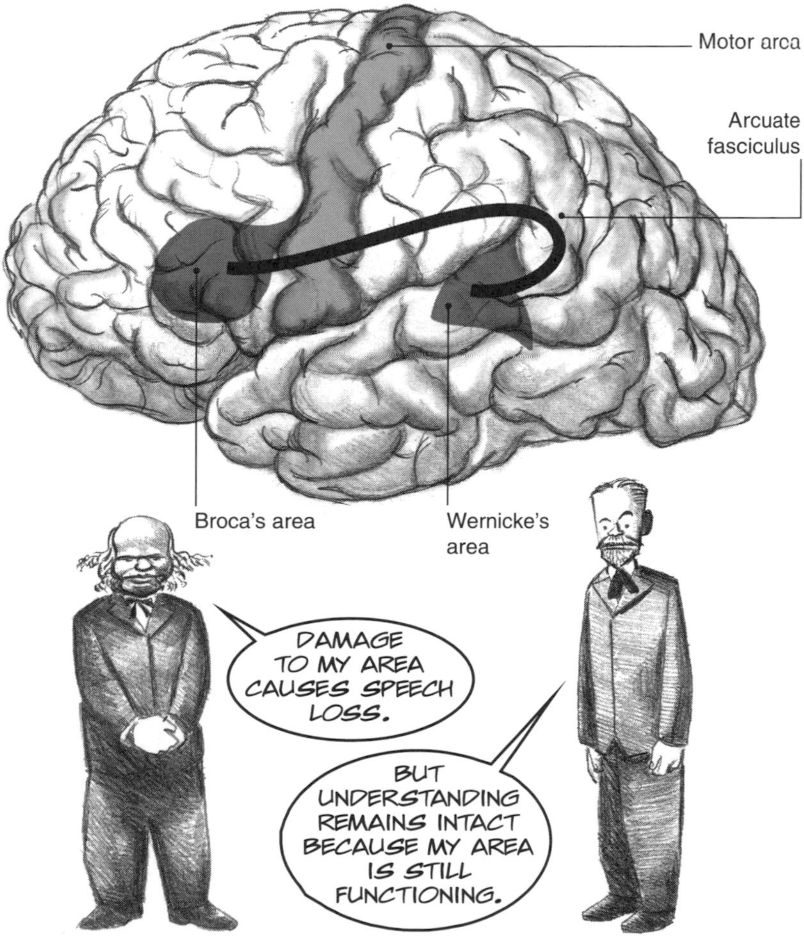
Motor arca
Arcuate fasciculus
Broca's area
Wernicke's area
DAMAGE TO MY AREA CAUSES SPEECH LOSS.
BUT UNDERSTANDING REMAINS INTACT BECAUSE MY AREA IS STILL FUNCTIONING.

In Wernicke's aphasia the person cannot translate between thought and language. She can still talk, because Broca's area still works, but what she says is mostly meaningless.

Wernicke's model is important because it explains various language disorders. It also shows that language involves the inter-play of many specialized brain areas. Language is too complex to be localized in just one centre.

However, even Wernicke's model is far too simple to explain all of language use. Modern investigators have repeatedly found that severe forms of the language disorders invariably involve **sub-cortical** as well as cortical damage. Once we realize that control of well-practised behaviours (habits) passes to sub-cortical centres, the reason for this becomes clear. Much of daily conversation is **routine**, and much of our speaking and listening is **inattentive**.

THINK HOW OFTEN WE TALK AND WRITE WITH STOCK PHRASES, EVEN WHEN TRYING HARD TO AVOID THEM.
"TRYING HARD" IS A STOCK PHRASE.
SO IS "STOCK PHRASE"!
"YES, DARLING, OF COURSE I'M LISTENING TO WHAT THE FOOTBALL COACH SAID."

Normal conversation only requires our attention intermittently. Life is too rich for attending to language all the time.

Language and *All* the Brain

Modern brain imaging allows us to study people as they engage in various language tasks. These studies confirm that the classical language zones in the LH (left hemisphere) are indeed active in speech and comprehension; but they also demonstrate that many other brain regions become active even in relatively simple tasks.

I SUGGESTED THAT THE RH MIGHT PLAY A FULLER ROLE IN LANGUAGE THAN WAS GENERALLY RECOGNIZED.
MODERN STUDIES DEMONSTRATE HOW CORRECT HUGHLINGS-JACKSON WAS. RH DAMAGE IN ADULTS CAN RESULT IN HESITATIONS AND REPETITIONS IN SPEECH.
SUCH PEOPLE MAY ALSO SPEAK IN AN EMOTIONLESS MONOTONE WHICH CAN BECOME VERY DISTURBING TO RELATIVES AND FRIENDS.
THEY ARE ALSO BAD AT RECOGNIZING EMOTION IN OTHER PEOPLE'S VOICES.
Have you got the time?
No, thanks.

RH damage also impairs understanding of many less obvious features of language, such as indirect questions (e.g. “Have you got the time?”, meaning something entirely different), sarcasm, humour and metaphor. These deficits reveal how complex language is – another important clue to the “mind”.

Language, Interpretation and Action

Read this statement:

“The lobster at eighteen is about to blow a fuse.”

At first, this may conjure up bizarre surrealist images. But imagine a busy restaurant with numbered tables – and one harassed waitress making that remark to another. Suddenly it makes sense.

Speech
“The lobster at eighteen is about to blow a fuse.”
UNDERSTANDING SPEECH IS NOT JUST ABOUT RECOGNIZING WORDS AND SENTENCES.
WE HAVE TO INTERPRET THEIR MEANING AND THE MEANING OF THE SPEAKER.

Speech is a form of action.

Speakers request, deny, cajole, inform, boast and so on. Listeners interpret what is said, and how it is said, in the light of their knowledge of the language, of the current physical and social context, and of the personality, intentions and quandaries of the speaker.

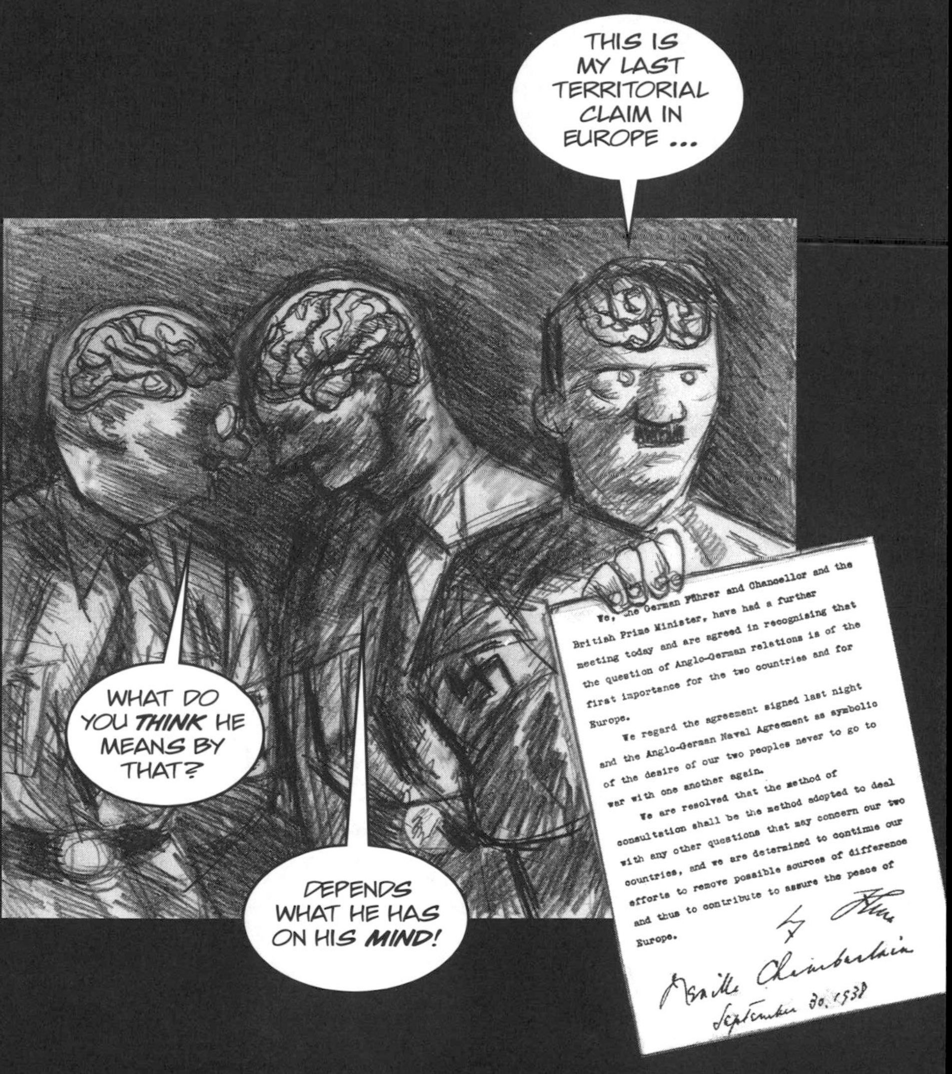

Both speaking and listening draw on all kinds of remembered information, on inferences, on projecting a certain image of oneself, and so on. Little wonder that normal language use involves areas *throughout the brain*.

Movement and Mind

The purpose of the brain is to produce *behaviour* – which is *movement*. Although we talk about **motor systems**, almost all of the brain is in some way involved in the control of movements. Even those areas that are supposedly dedicated to sensation. For example, it is hard to walk when your leg has "gone to sleep". Without sensory feedback on how "they are doing", the motor systems do not do very well.

Tuning the Movements

In both evolution and individual development, control of movement spreads outwards from the body to the limbs, and along the limbs to the digits. The baby in the womb makes whole body movements. Soon after birth her limbs make gross flailing movements.

Within weeks, she has sufficient control to scoop up objects with an arm.

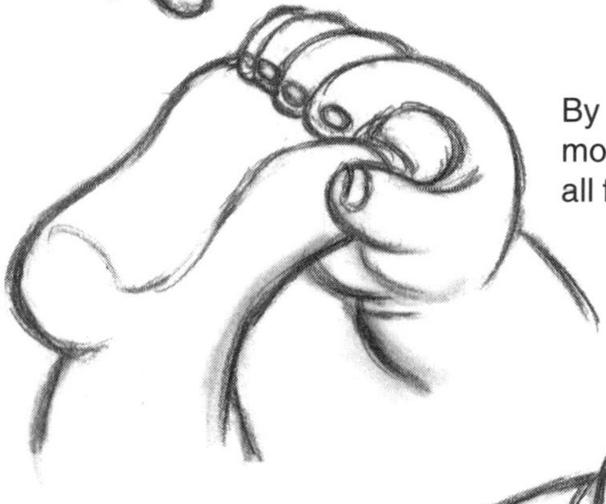

By the second to fourth month, she can **grasp** with all fingers at once.

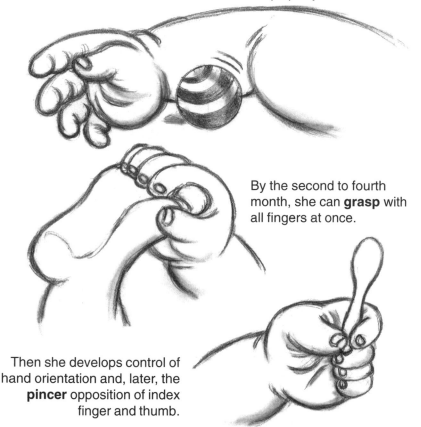

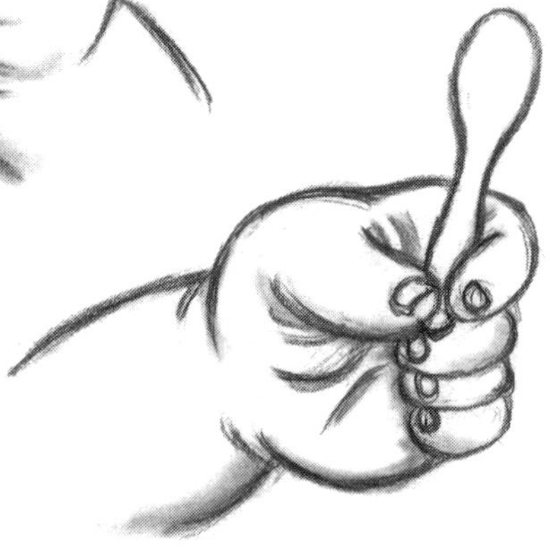

Then she develops control of hand orientation and, later, the **pincer** opposition of index finger and thumb.

The development from gross to fine movements follows a principle of **inhibitory tuning**. Fine movements rely on the same commands as gross movements, but narrow their range of application. You can see this by trying to bend one of your fingers while keeping all the others straight. For the index finger this is not too difficult. It gets harder with the fingers you least often use for voluntary actions. It is inhibitory tuning that gradually "sculpts" the gross flailings of the infant into finely controlled actions.

Two Motor Movement Control Systems

Picking up an object involves two components:

These components are controlled by separate motor fibres running from brain to spine: the **extrapyramidal** fibre tract and the **pyramidal** fibre tract.

Lesions to one or other impair the corresponding movement component.

For example, damage to the **descending pyramidal** tract reduces the efficiency of **grasping** but has little effect on the timing or accuracy of **reaching**.

Levels of Control of Movement

Control of movement illustrates the notion of **levels of control**. At the lowest level is **spinal control**. This covers reflexes (e.g. the knee jerk), which maintain muscle tone and posture, and spinal programming of **movement patterns** such as upright walking.

In between these two extremes are many gradations of automaticity and compulsion. Normal breathing is unlearnt and largely automatic, whereas walking is learnt with difficulty but becomes semi-automatic. Compulsions include tics, the needs to stretch and yawn, and various urges to touch. Let's now see how these gradations originate in the motor system.

The Motor System

Gradations of automaticity reflect the levels of control in the motor system: **spine**, **brainstem**, **cerebellum**, **basal ganglia** and **cortical motor areas**.

Major structures and connections in the organization of the motor system

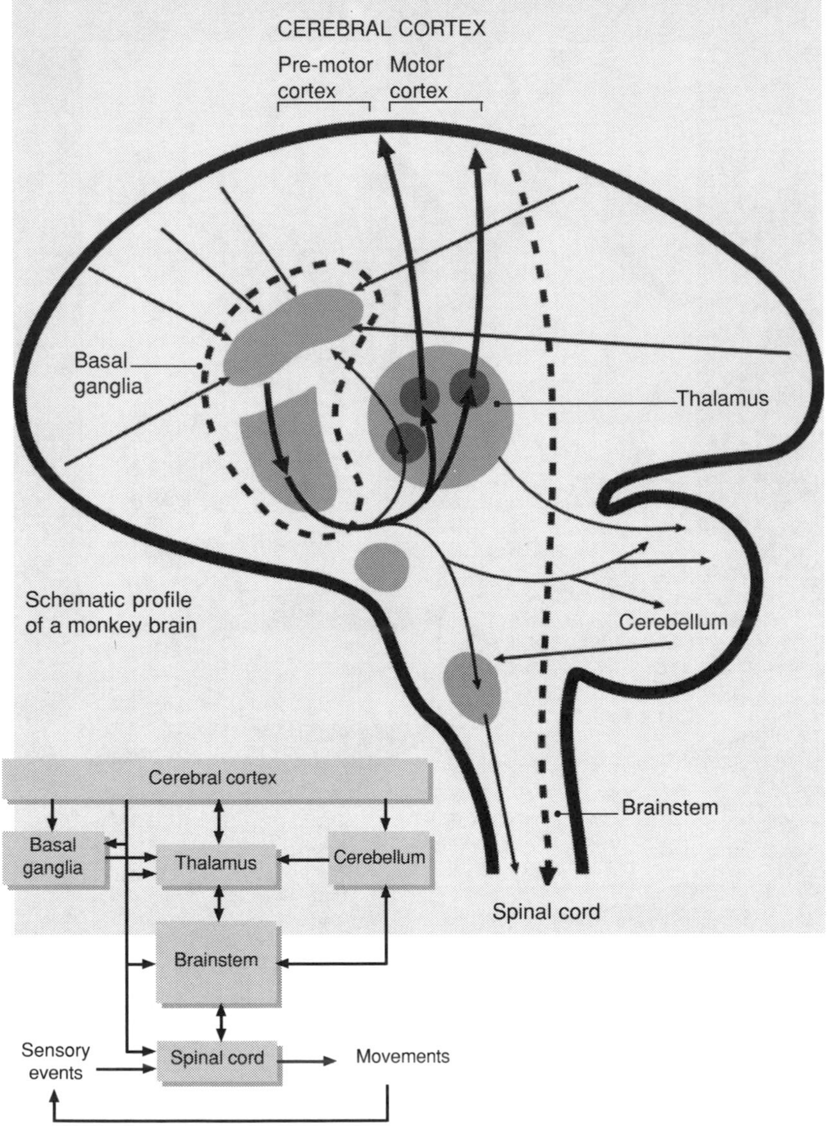

Damage to the Motor System

Wherever they originate, all forms of movement are finally expressed as the firing of **motor neurones** in the **brainstem** and **spinal column**. Destruction of these results in paralysis of corresponding body parts.

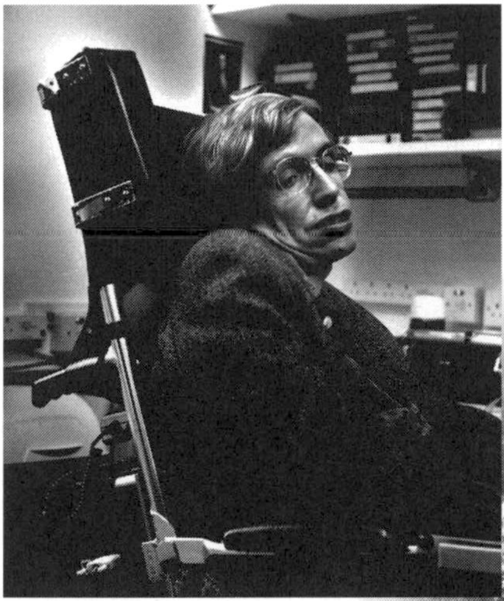

The world-famous cosmologist Stephen Hawking suffers from Motor Neurone Disease

The "movement melody" of the healthy person

At the next level of control is the **cerebellum**. Damage here has a variety of consequences. Defects include loss of the ability to learn new movements, disruption of posture, jerkiness of movements, inability to make rhythmic movements, and impaired sequencing of movements. The cerebellum seems to fulfil several roles. It stores skilled movement sequences, adds fine tuning and timing to movements selected elsewhere, and composes them into the **movement melody** of the healthy individual.

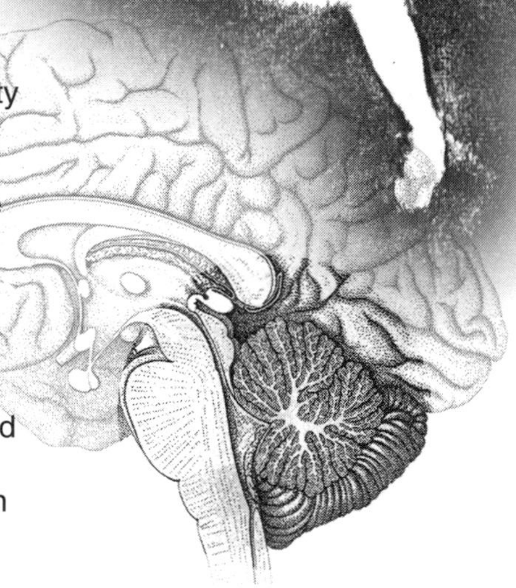

The functions of the **basal ganglia** (BG) are just as complex as those of the cerebellum. People with Parkinson's disease, characterized by tremor and an inability to initiate movements, have a shortage of dopamine in the BG. Abnormalities in the BG also accompany Huntington's disease, a degenerative condition with symptoms including involuntary grimacing, twitching and body twisting.

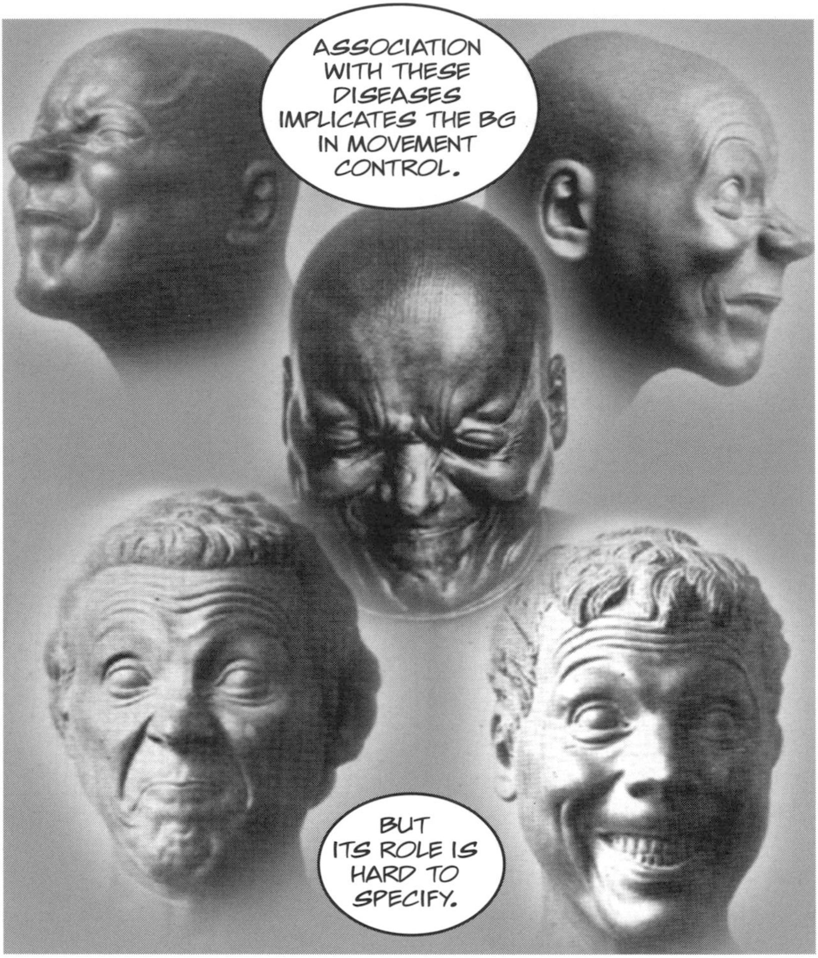

One theory has the BG responsible for the force, direction, extent and duration of movements. Errors in computing the force needed to make a movement could take the form of a failure of initiation, as in Parkinsonism. Or they could lead to excessive initiation followed by a series of overcompensations, producing the grotesque agitation of the Huntington's sufferer.

Destruction of the highest motor centre, the **primary motor cortex**, results in loss of skilled and delicate movements, particularly of the hands and fingers. This is because the **pyramidal** fibres, which control the hands, originate in the motor cortex.

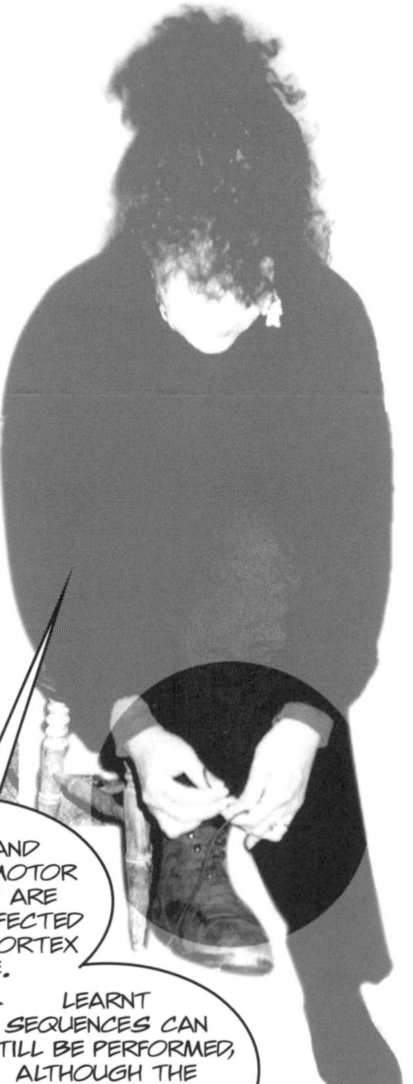

BOTH LEARNING AND MEMORY OF MOTOR SEQUENCES ARE LARGELY UNAFFECTED BY MOTOR CORTEX DAMAGE.
LEARNT SEQUENCES CAN STILL BE PERFORMED, ALTHOUGH THE EXECUTION IS SOMEWHAT CLUMSY.

Presumably, movement *learning* and movement *memory* are being handled in the cerebellum.

The Origins of Voluntary Movement

Damage to the rear of the **left parietal lobe** gives rise to a condition of **ideomotor apraxia**. People have difficulty making movements and gestures. The problem is less severe for the use of concrete objects ("Show me how to use a hammer."), especially if the object is actually present.

It is most severe for symbolic gestures, such as greeting or saluting. Particularly when these have to be performed outside their normal social context.

What is lost is the ability to make voluntary movements that are not prompted by the environment.

The left parietal lobe may play a role in voluntary movement because it is close to the language centres.

According to **Lev Vygotsky** (1896-1934), voluntary action starts as something shared between a child and an adult. The two attend to the same object and the adult gives instructions which the child learns to obey.

Subsequently, as she learns to speak, the child uses the same spoken commands to control her own behaviour. Eavesdropping on a lone three- or four-year-old reveals lengthy bouts of self-instruction. With age, however, self-directing speech becomes internalized. (Although this may be the case more for literate cultures in which talking to yourself is not judged kindly!)

Proprioception and Body Ego

Because movement control happens at many levels, the motor system is forgiving of damage in any one place. Intact structures are always capable of some residual movement. It is ironic that one of the most devastating losses of movement results from a **sensory** defect.

PROPRIOCEPTION IS THE LARGELY UNCONSCIOUS SENSE OF WHERE OUR BODY PARTS ARE IN SPACE.
Muscle fibres
Incoming nerve
Outgoing nerve to muscle
Proprioceptive (stretch) receptors are in the muscles and joints
NERVE CELLS IN THE MUSCLES AND TENDONS SIGNAL THE DEGREE OF STRETCH OF OUR MUSCLES AND JOINTS.

"**Who am I?**" is also a bodily question of "**Where am I?**"

Occasionally, disease or vitamin overdose extinguishes proprioception. This produces a total loss of the **body sense** and, with it, the **body-ego**. The person feels disembodied and therefore cannot generate movements. Loss of body sense teaches another important lesson about the link between *movement* and *mind*.

Smells and Emotions

The **limbic system**, sometimes known as the **emotional brain**, has a major role in the experience and expression of emotions. The limbic system initially evolved for evaluating smells.

Some of the major elements of the limbic system

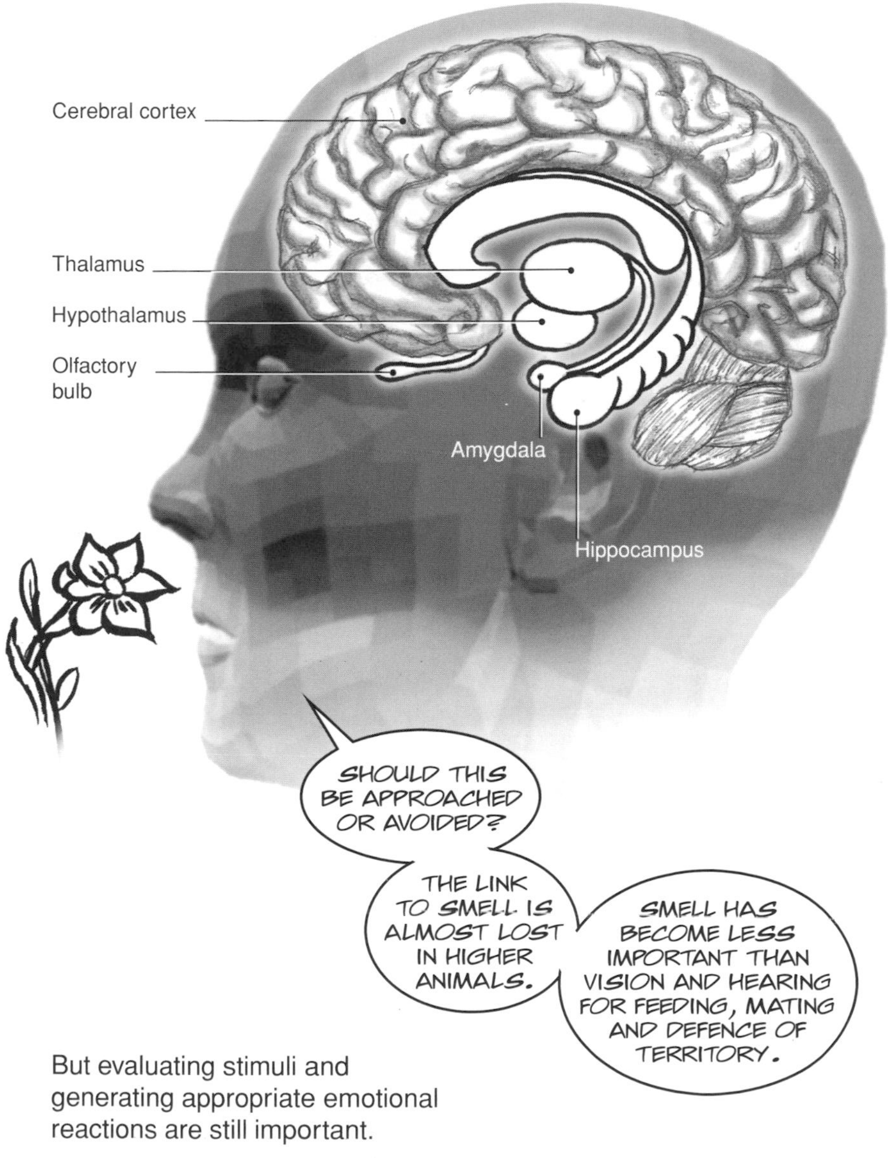

Cerebral cortex
Thalamus
Hypothalamus
Olfactory bulb
Amygdala
Hippocampus
SHOULD THIS BE APPROACHED OR AVOIDED?
THE LINK TO SMELL IS ALMOST LOST IN HIGHER ANIMALS.
SMELL HAS BECOME LESS IMPORTANT THAN VISION AND HEARING FOR FEEDING, MATING AND DEFENCE OF TERRITORY.

But evaluating stimuli and generating appropriate emotional reactions are still important.

Emotional Reaction

When you are joyful or angry, your limbic system is active. **Epileptic seizures** confined within the limbic system produce strong emotional reactions, ranging from terror to elation.

Limbic stimulation with electrodes produces emotional displays in animals. Whereas damage to the system leads to loss of normal emotional behaviour.

Emotions are complex and can involve many brain areas besides the limbic system. Studies of **fear** illustrate this.

The Anatomy of Fear

If an animal learns to press a lever to get food and then receives an **electric shock**, two things happen. The animal's heart rate shoots up, and it ignores the lever for a while. These are two measures of **unlearnt fear**.

If, next, a **tone** is paired with the shock for several trials, then when the tone sounds alone it will cause both the heart rate increase and the suppression of lever pressing. Here, the two measures show **learnt** (or **conditioned**) **fear** of the tone.

Fearful Symmetry

If a tiny lesion is now made in a certain region of the animal's **hypothalamus**, its heart rate no longer shoots up when the tone sounds, but it still stops pressing the lever. The lesion abolishes one expression of learnt fear but not the other. Yet if the animal receives another shock, without the tone, then it still shows both unlearnt heart rate change and unlearnt suppression of lever pressing.

And for heart-rate change, different circuits carry the learnt and the unlearnt fears.

This may seem complicated. It **is** complicated. It is also characteristic of the complex relationships between brain and behaviour – or brain and *mind*. We will encounter many more examples of this sort. Here is another one relating to the emotion of fear.

Sub-cortical Learning

Information from the eyes and ears travels first to the **thalamus**, and from there to the visual and auditory areas of the cortex. It used to be thought that sights and sounds were **first** experienced and recognized in these cortical areas. Information about what had been recognized was **then** supposedly sent to the limbic system for an emotional reaction: "Is it good or bad?"

It has been discovered, however, that in addition to this indirect route (thalamus→cortex→amygdala), there is a direct route from the thalamus to the amygdala.

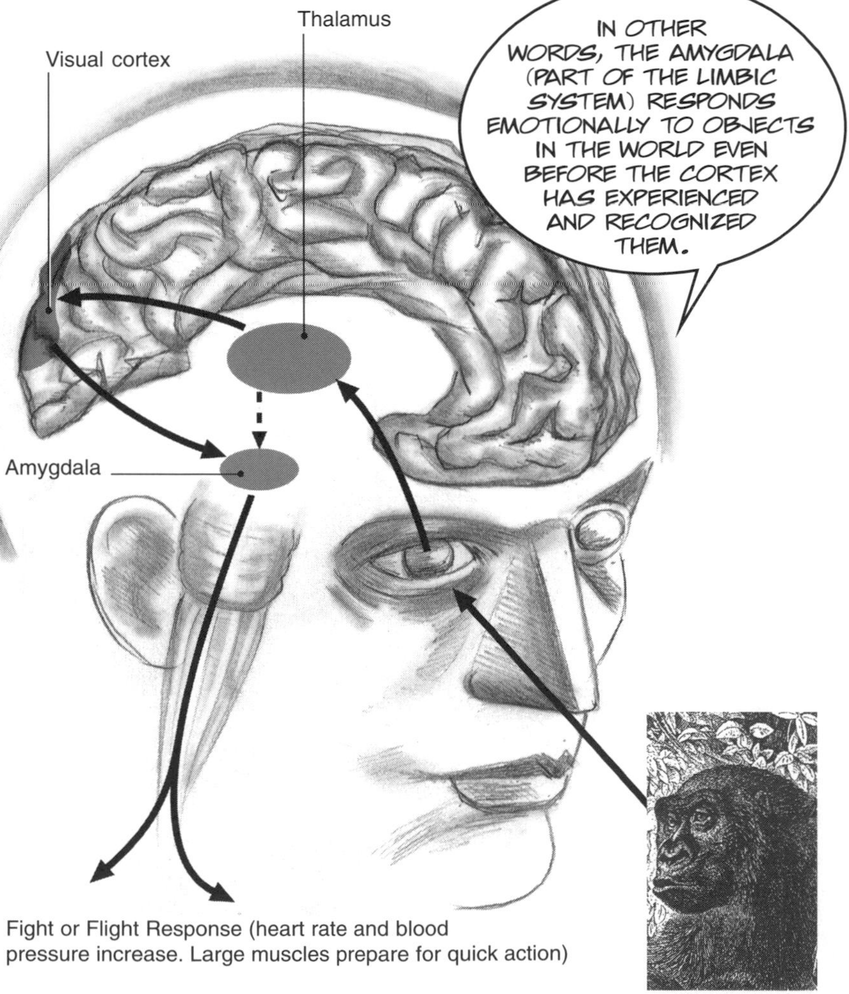

Thalamus
Visual cortex
IN OTHER WORDS, THE AMYGDALA (PART OF THE LIMBIC SYSTEM) RESPONDS EMOTIONALLY TO OBJECTS IN THE WORLD EVEN BEFORE THE CORTEX HAS EXPERIENCED AND RECOGNIZED THEM.
Amygdala
Fight or Flight Response (heart rate and blood pressure increase. Large muscles prepare for quick action)

Knowing When to be Afraid

If rats with their auditory cortex removed experience a tone paired with a shock, they quickly learn to fear the tone.

The amygdala and other limbic structures perceive, remember and learn, as they presumably do in lower animals that have no cerebral cortex.

Recall the herring gull chicks apparently pleading for food. Their behaviours probably arise in something of the same fashion. They have brain circuits that respond to the simple feature **red dot on yellow**, not to the complex shape of an adult bird.

Similarly, many animals show freezing and fleeing responses to the movement of passing clouds and swaying branches. The circuits are there to detect the movements of potential predators. They are easily triggered by inappropriate stimuli.

So do humans also show emotional learning without conscious cognitive involvement? This could explain why our emotional responses sometimes seem uncalled for. A strong emotional response to a stranger may be a learnt reaction to some feature the stranger shares with a person we knew before.

Emotions "Left and Right"

It would be wrong to suppose that only the **limbic** system plays a part in emotions. After all, we sometimes have strong emotional reactions only after we have used our **neo-cortex** to **consciously** think through a turn of events or a conversation.

Look at the cartoon faces below. Focus on the nose of each in turn, and decide which one looks happier.

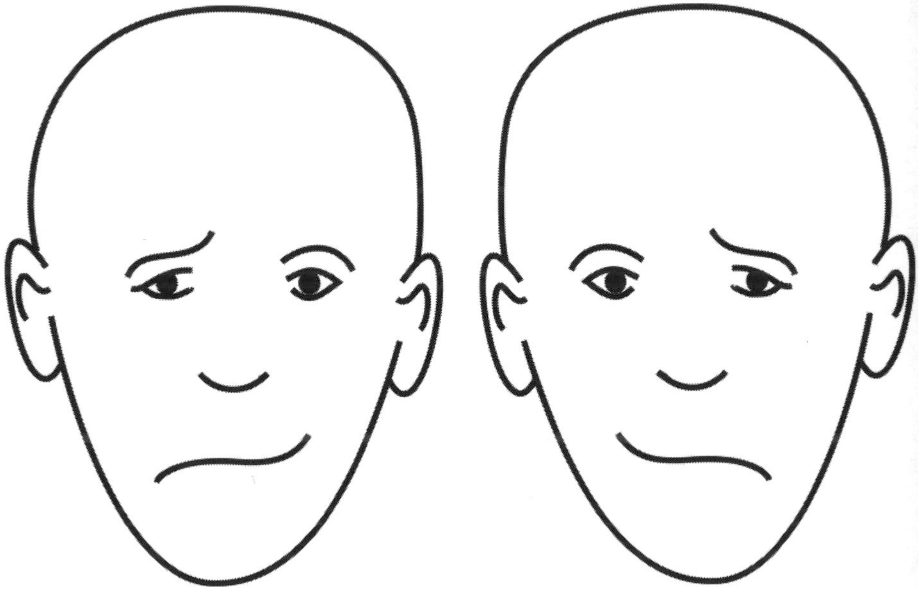

Although they are mirror images of each other, most people say the face on the right looks happier.

This is because the **left** half of each face is being seen first by the **right hemisphere** (RH) of your brain, which is specialized for face processing. Your judgement of the emotions on each face is driven more by information from the left of each image than by information from the right.

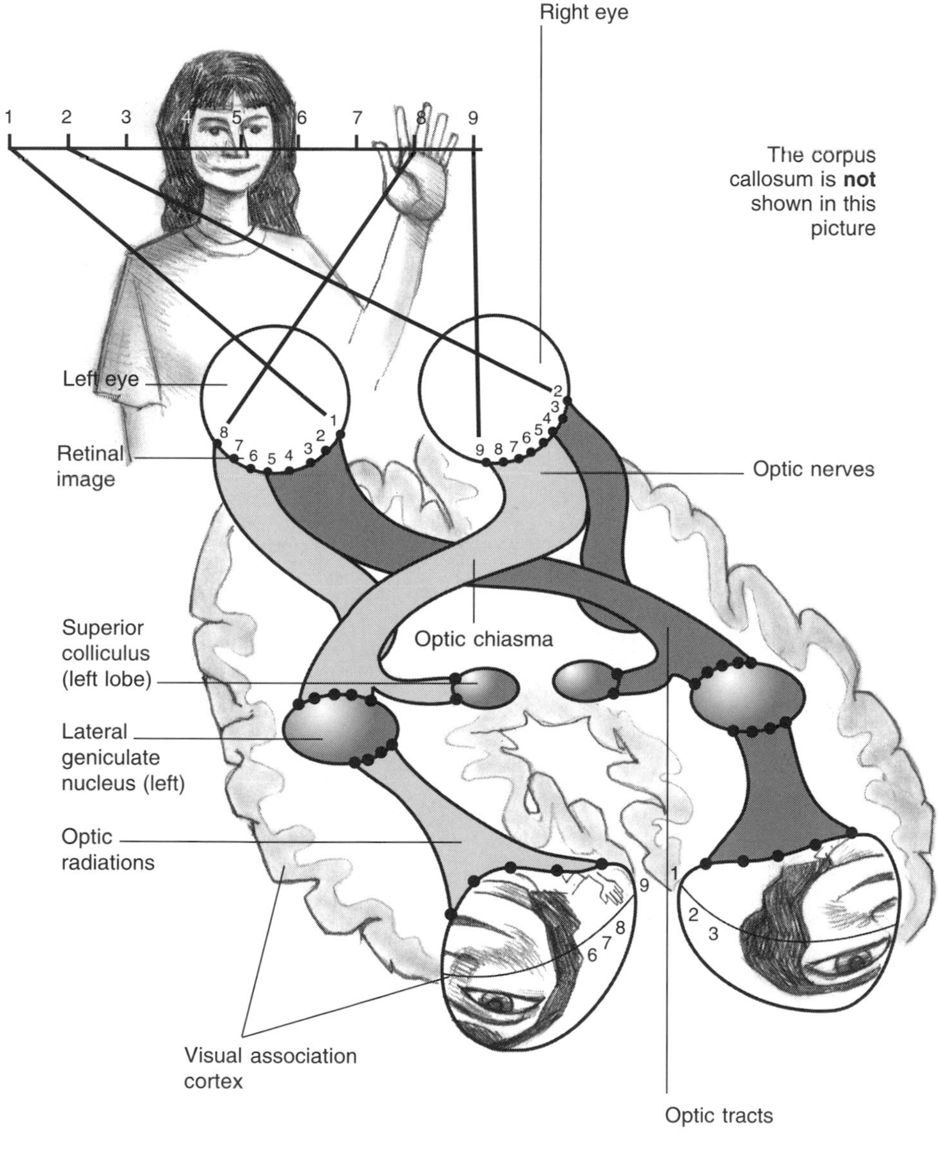

Emotional Tone

The RH also plays a bigger role than the LH in judging the emotional tone of voices. People with Wernicke's aphasia from LH damage no longer understand language. Yet they judge a speaker's emotional tone better than **normal** people or those with RH damage.

There are also hemispheric differences in the generation of emotions. The LH seems connected with more positive emotions than is the RH. People with LH damage tend to develop **depression**, whereas those with RH damage are prone to **manic cheerfulness**. In each case the undamaged hemisphere, no longer restrained by its twin, shows its true emotional colours.

Emotion and Reason

Emotions have sometimes been thought rather an intellectual embarrassment – an inheritance of our "animal nature".

EMOTIONS ARE STATES OF THE BODY, WHEREAS THOUGHTS ARE STATES OF THE MIND OR SOUL.
EMOTIONALITY IS SEEN AS A MORE PRIMITIVE MEANS FOR DEALING WITH THE WORLD THAN RATIONALITY.
EMOTIONS NEED TO BE SUPPRESSED IN THE INTERESTS OF ATTAINING PURE RATIONALITY.

Plato (428-348 BC)

Immanuel Kant (1724-1804)

But unless the rational mind is a divine gift – something over and above our biological nature – this will not do. Thought and emotion are both expressions of *brain activity* and must be as mutually interdependent as any bodily functions.

Emotions Involved in Decisions

The **limbic system** has strong connections to the **frontal lobes**. When these connections are damaged, people may show surprisingly little intellectual impairment. Yet their personal, social and professional lives fall apart. The problem lies in their *decision-making*. Faced with a problem requiring a decision, they analyze and evaluate all the alternatives, often at excessive length, and might finally make their choice for irrelevant reasons. Take the example of a patient asked on which date to see his doctor …

They can talk rationally, distinguish what is socially acceptable or not, but they do not seem to **feel** their own emotional assessments at a gut level. They might even remark that although they **know** what they should be feeling, they do not consciously **have** the feelings.

Study of such people has made it clear that emotions are an important part of normal reasoning and decision-making. When a normal person faces a problem, he or she simply will not bother to consider many of the possible solutions. Only possible solutions that have the "right feel" about them get chosen for conscious consideration.

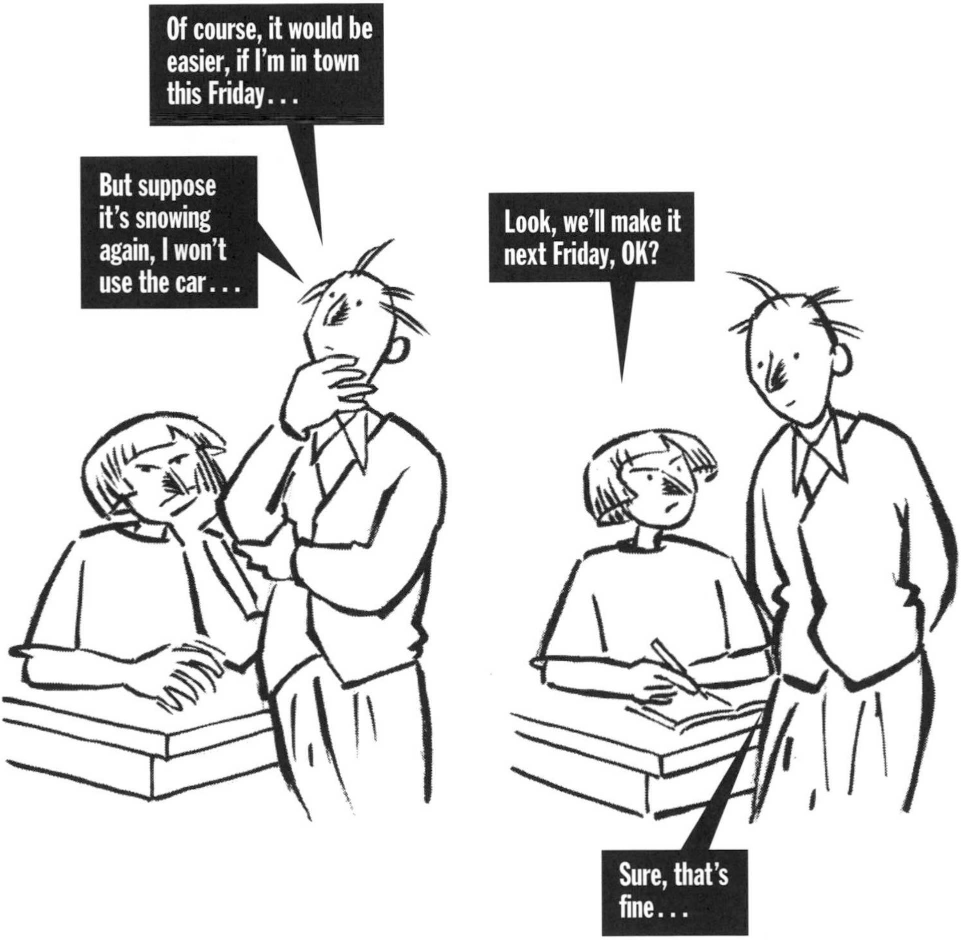

Trivial problems are not endlessly analyzed, because they are just not worth the cost of prolonged soul-searching. People with damage to those areas of the frontal lobes that receive limbic input appear to lose this emotional guidance of their thought processes.

Memory Makes You Flexible

Emotions may serve to guide reasoning, but originally they must have guided automatic behaviours, making them more flexible. A non-specific emotional reaction, such as a startle response, can serve a general arousing function, preparing an animal for action of some kind.

Consider the rat that hears a tone shortly before it receives a shock. The shock produces unlearnt fear, and through **conditioning** the tone comes to provoke learnt fear. Now when the rat hears the tone, it wants to escape. Its behaviour is more flexible because it no longer has to wait for the actual shock to "know what to do".

This kind of learning is especially important for animals that explore the world through smell. They detect potential food, mates and predators at a distance, often long before making visual contact. This means they can start seeking out or escaping from the source of the smell in good time. If they are also capable of emotion-based conditioning, they can acquire a wide repertoire of approach and avoidance responses. This allows more flexible behaviour than if all their responses were "wired in" at birth.

For the novelist, **Marcel Proust** (1871-1922), the taste of a particular tea and cake opened up the memory of an entire past.

It is not surprising, therefore, that close by the limbic system, which started out as a "smell brain" and evolved into an "emotional brain", is an area of cortex that is important in learning and memory. This is the **rhinal cortex** on the lower inner surface of the **temporal lobes**.

What Amnesia tells us about Mind

Damage to the rhinal cortex of both hemispheres causes severe memory loss or **amnesia**. The crucial feature of the **amnesic syndrome** is a devastating absence of memory for events since the injury (**retrograde** amnesia).

Amnesics can appear perfectly normal on brief acquaintance, but not for long. They forget information and events within minutes.

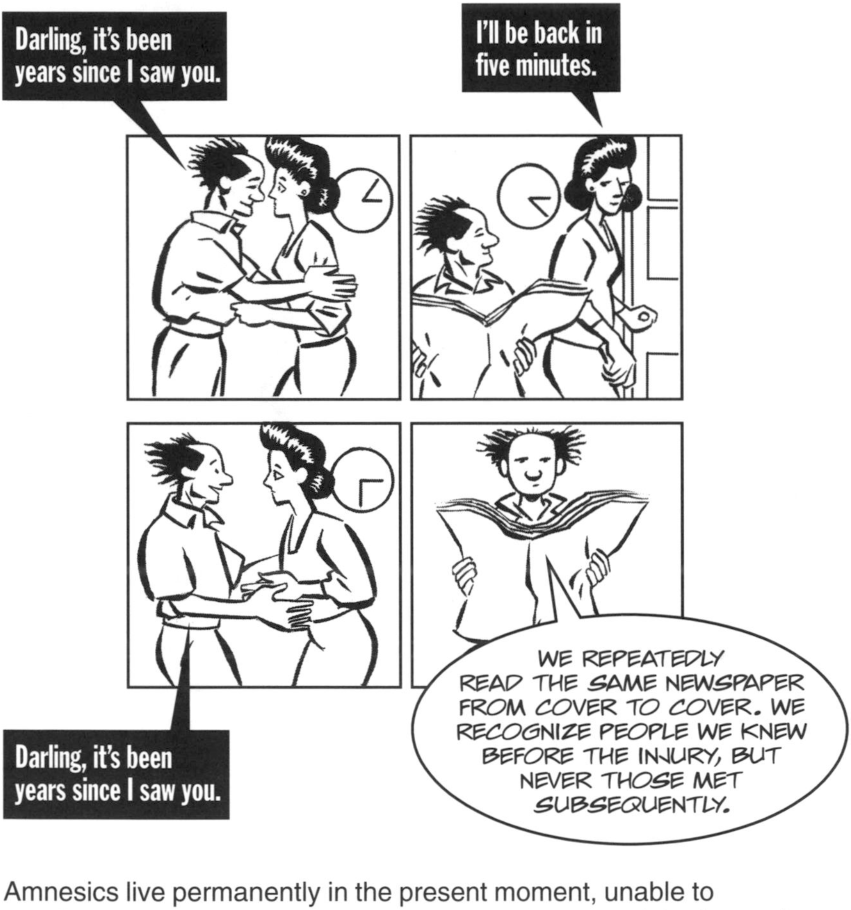

Amnesics live permanently in the present moment, unable to recollect either their recent past or their anticipations of the future. It is as if they are in a permanent state of having just woken up.

N.B. Although there are individuals who forget "who they are", this is *not* the usual meaning of "amnesia".

Two Kinds of Memory

Because amnesics can recall long past events but not recent ones, it suggests that the rhinal cortex must be involved in the **storage** of new memories rather than in memory **retrieval**. However, even densely amnesic patients can store certain sorts of new memory. This applies to **procedural** (how to) skills, such as typing or roller-blading. Amnesics may do almost as well as normal people at acquiring new procedural skills.

They also show normal **perceptual learning** and **memory**.

Examples of perceptual learning are learning to identify species of flowers or birds, recognize when pastry has the proper consistency, or hear if the timing of an engine sounds right. Laboratory demonstrations of perceptual learning often involve puzzle pictures, like the one shown above. Can you see what it is?

Memory with and without Emotions

Like many images, such as X-ray images, puzzle pictures have to be interpreted. Once people have learnt to see such pictures "correctly", they never forget "how to do it". Amnesics perform just the same, although on re-testing after a few hours or days they deny ever having seen the pictures before.

It seems, therefore, that the rhinal cortex processes memory for new **episodes** that have been experienced but not memory for new **know how** procedures. This seems logical.

- Episodes in our lives give rise to *emotions*.
- The limbic system is crucial in *emotional experience* and is next to the rhinal cortex.
- The rhinal cortex is important for *memory of life episodes*.

It makes sense to remember events that are emotionally arousing because they are likely to be important to us. For this reason, the same neurochemicals whose release into the bloodstream puts the body on alert also instruct the brain to store a lasting record of the moment.

By contrast with memory for **autobiographical episodes**, procedural (how to) memories are not emotionally loaded. Although we take pleasure in successfully exercising our procedural skills, or are frustrated by our failures, these emotions attach to *individual episodes* of skill use, rather than to the procedural skill itself.

Animals evolved motor skills memory long before emotions arrived on the scene. Think of Aplysia which is capable of habituation and sensitization. Such examples suggest that memory for motor skills will be lodged in relatively old, low-level brain structures. And this turns out to be the case.

The Location of Memories

An example comes from eyeblink conditioning in rabbits. A puff of air (UCS) to the eye causes a reflex blink (UCR). If the puff is paired with a tone (CS) over many trials, then eventually conditioned blinks (CR) occur in response to the tone alone. A tiny lesion in the cerebellum abolishes the conditioned blink, but leaves the reflex blink to the air puff unaffected. The **memory trace** for the conditioned blink is in the cerebellum.

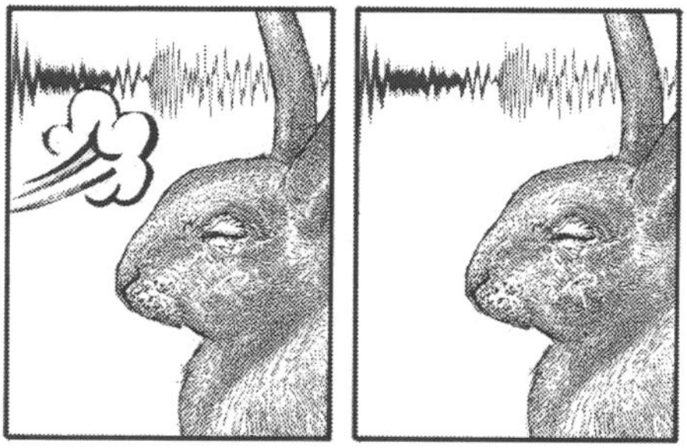

Amnesics also show eyeblink conditioning. If the pairings of tone and air puff occur on one day, and testing with the tone alone on the next day, the amnesic shows the conditioned blink response to the tone but denies any memory of the conditioning trials. People with cerebellar damage, by contrast, can recollect the conditioning trials but never acquire the conditioned blink!

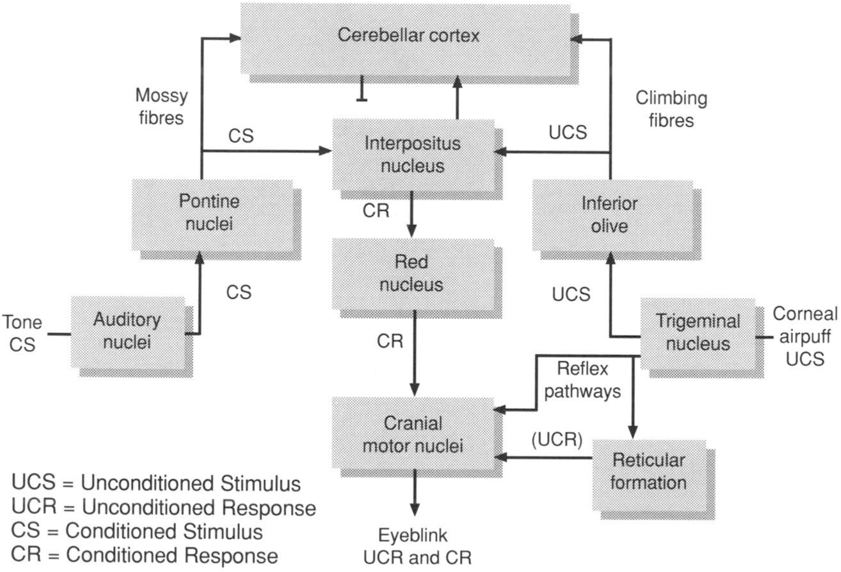

In the 1930s, the neuropsychologist **Karl Lashley** (1890-1958) tried to locate the seat of memory by training rats on simple tasks and then removing different portions of their brains.

I FOUND THAT THE MORE TISSUE I REMOVED, THE WORSE THEIR PERFORMANCE BECAME.
BUT THERE WAS NO ONE SITE WHOSE REMOVAL TOTALLY ERADICATED MEMORY FOR A PARTICULAR TASK.

These results led Lashley to adopt a wholistic view of brain function. He was correct that there is no single site of memory, but wrong about wholism. Memories **do** reside in specific circuits, sometimes even in particular components of a circuit. But memories are much more complex than was once realized, as we shall see next.

The Complexity of Memory

For example, a chick will peck at a shiny bead. Coat the bead with a bad-tasting liquid and the chick no longer pecks at it. The chick has developed an aversion. This looks as if it should be a single memory. Yet it turns out the chick has actually learnt *three* aversions: to the shape, the taste and the shine of the bead.

EACH MEMORY IS IN A DIFFERENT PART OF MY BRAIN.
NO WONDER LASHLEY FAILED TO FIND A SINGLE SITE FOR MEMORY IN ANIMALS LIKE ME WITH LARGER BRAINS THAN THE CHICK.

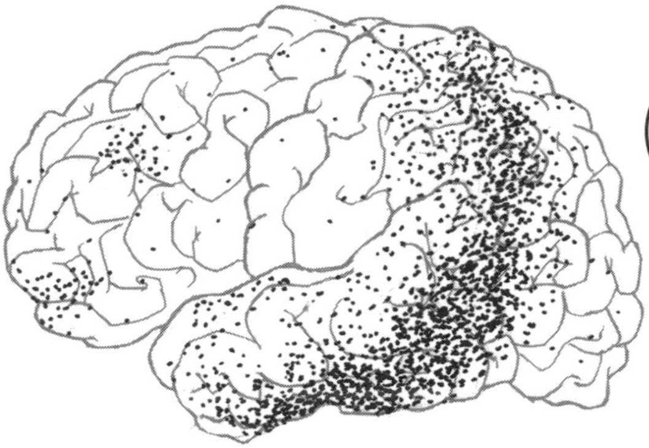

Schematic representation of the distribution and severity of degeneration of the brain in an average Alzheimer case. The darker the area, the more pronounced the degeneration.

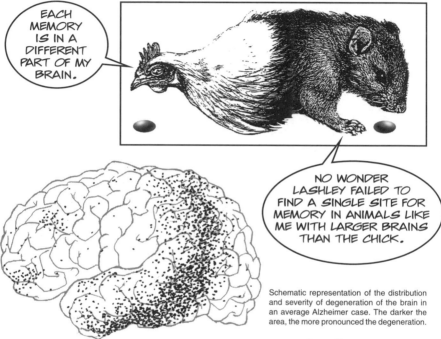

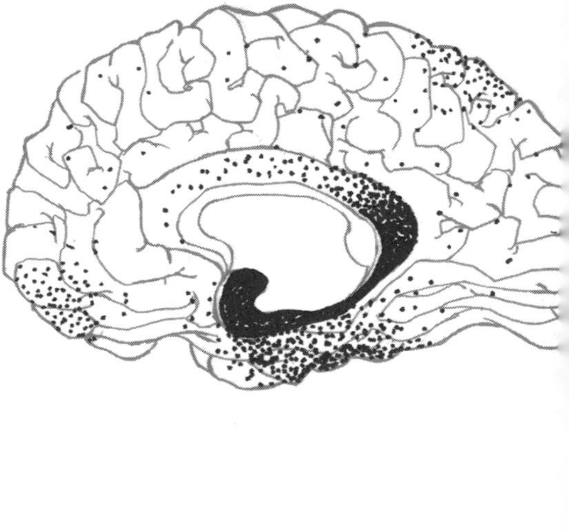

Memory loss is a prominent symptom of **Alzheimer's disease**. Cell death in Alzheimer brains is particularly severe in the region of the rhinal cortex, but there is also extensive temporal and parietal lobe degeneration. No wonder, then, that Alzheimer patients exhibit both the symptoms of the amnesic syndrome and a range of other memory problems.

Sensing and Seeing

Like other animals, humans learn about the world through their senses. Traditionally, there were **five senses**. Taste and smell are closely associated with the limbic system, situated deep in the brain. Vision, hearing and the touch senses are all heavily represented in the cortices (although they also connect to lower brain structures). The regions at which information from the senses first arrives in the cortex are the **primary sensory areas**.

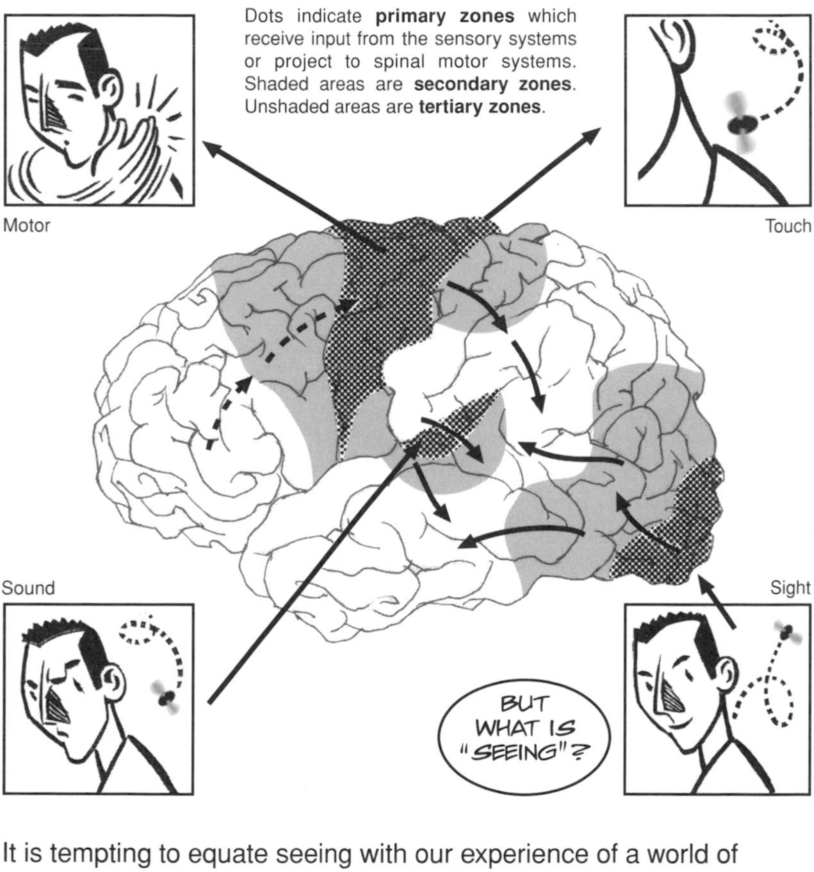

Dots indicate **primary zones** which receive input from the sensory systems or project to spinal motor systems. Shaded areas are **secondary zones**. Unshaded areas are **tertiary zones**.
Motor
Touch
Sound
Sight
BUT WHAT IS "SEEING"?

It is tempting to equate seeing with our experience of a world of familiar objects with specific locations and colours. This is seeing of a very high order.

No other animal has as much visual knowledge of the world as we have, because none has so much cortex dedicated to analysis of the information in light.

The Anatomy of Vision

At its simplest, seeing is merely the registering of light and some reaction to it. Many creatures that live under rocks show light avoidance responses. Our own visual systems include low-level functions. There are seven known pathways from the retina to the brain. The pathways to the **pineal gland** and the **superoptic nucleus** regulate bodily rhythms in response to the daily cycle of light and dark. The rest of our high-performance visual system evolved through additions to such humble beginnings.

The rest of this section on vision deals solely with the major pathway from retina to primary visual cortex (also known as **visual area 1** [V1] and other names besides). It contains many times more axons than all the other pathways put together and has its own component strands.

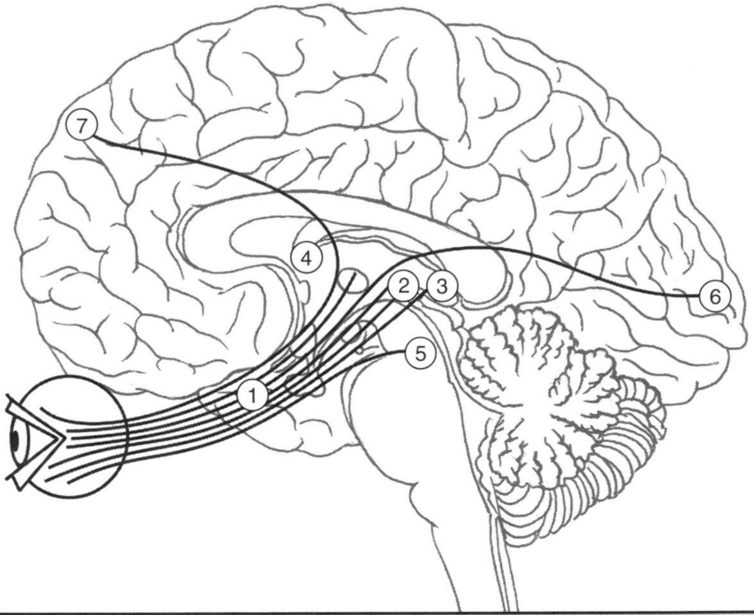

	Visual system	Postulated function
1.	Superoptic nucleus	Controls daily rhythms (sleep, feeding, etc.) in response to day-night cycles
2.	Pretectum	Produces changes in pupil size in response to light-intensity changes
3.	Superior colliculus	Head orienting, particularly to objects in peripheral visual fields
4.	Pineal body	Long-term circadian rhythms
5.	Accessory optic nucleus	Moves eyes to compensate for head movements
6.	Visual cortex	Pattern, perception, depth perception, colour vision, tracking moving objects
7.	Frontal eye fields	Voluntary eye movements

Each half of the visual field connects to V1 of the **opposite hemisphere**. In normal brains the LH and RH share information about the two halves of the visual field by means of the huge bundle of fibres known as the **corpus callosum**.

Information from the retina travels via a part of the thalamus called the **lateral geniculate nucleus** (LGN) to the primary visual cortex, V1. Points that are next to each other on the retina connect to cells next to each other in V1, and damage to V1 causes a blind spot (**scotoma**). Cells in V1 also connect back to the LGN, and this **two way neural traffic** is characteristic of the visual system and of the brain as a whole.

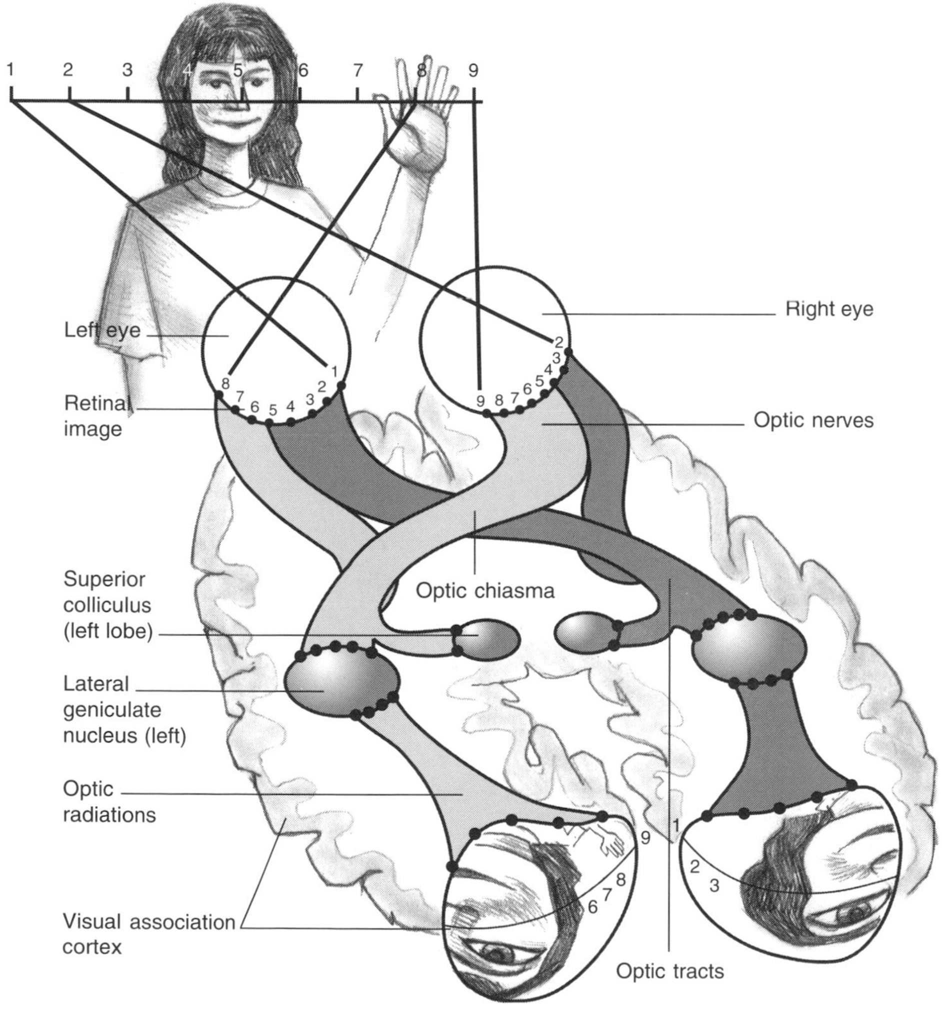

Visual Areas: Colours, Directions and Shapes

V1 is only the first of several "early" visual areas in the **occipital lobe**. Cells in V1 connect to cells in V2, which then project to a number of visual areas known as V3, V3A, V4 and V5. Cells in V4 increase their firing in response to **specific colours**, while those in V5 respond to objects moving in particular **directions**. Cells in V3 and V3A respond to lines at **particular orientations** (vertical, 5° clockwise, 10° clockwise, etc.) which allows them to analyze **shape**.

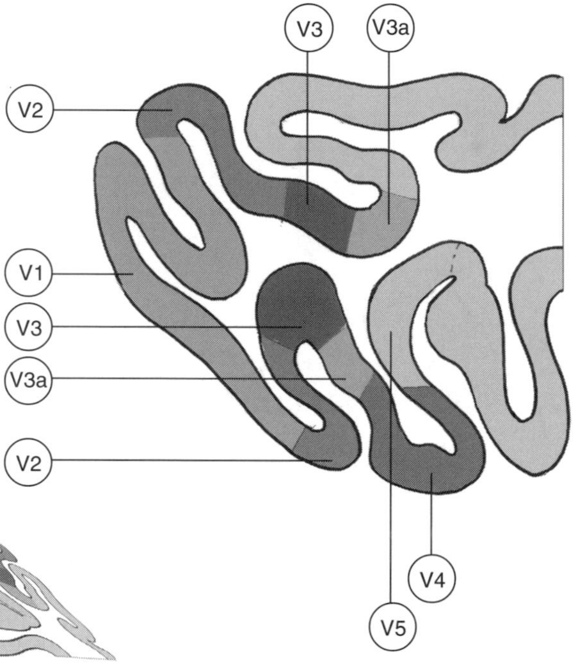

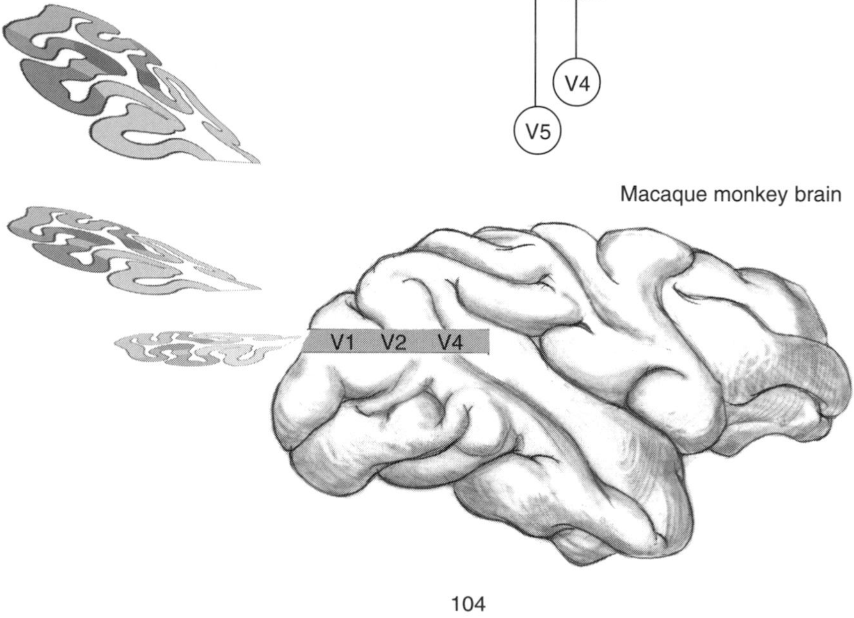

Macaque monkey brain

Loss of Colour

Brain imaging studies show V4 becomes active when people view coloured patterns and that moving displays activate V5. In addition, damage to V4 causes a loss of colour vision. This is **achromatopsia**: it is different from ordinary colour blindness.

If the V4 of only one hemisphere suffers damage (**unilateral** damage), then the opposite half of the world appears in black and white …

… and the same side half still in colour.

When the damage is **bilateral,** the patient becomes not only totally colour blind but also unable to remember or imagine colours. Colour no longer exists as a **category of experience**.

Motion Blindness

Damage to V5 produces the bizarre condition of "motion blindness". The person still sees forms and colours but the experience of moving objects becomes somewhat like viewing a series of still photographs. An approaching object grows larger and closer in discrete jumps, making it difficult, for example, to cross a road safely.

Higher Level Vision

Only the very first processes of vision take place in the **occipital lobes**. The **temporal**, **parietal** and **frontal** lobes also contain many areas engaged in processes related to vision. In fact, it requires considerable courage even to look at a representation of all the known visual areas and their interconnections.

Three main visual pathways leave the occipital lobes. They connect to the temporal lobe (lower pathway), the upper temporal sulcus (middle pathway), and the posterior parietal lobe (upper pathway). Each stream processes certain types of visual information.

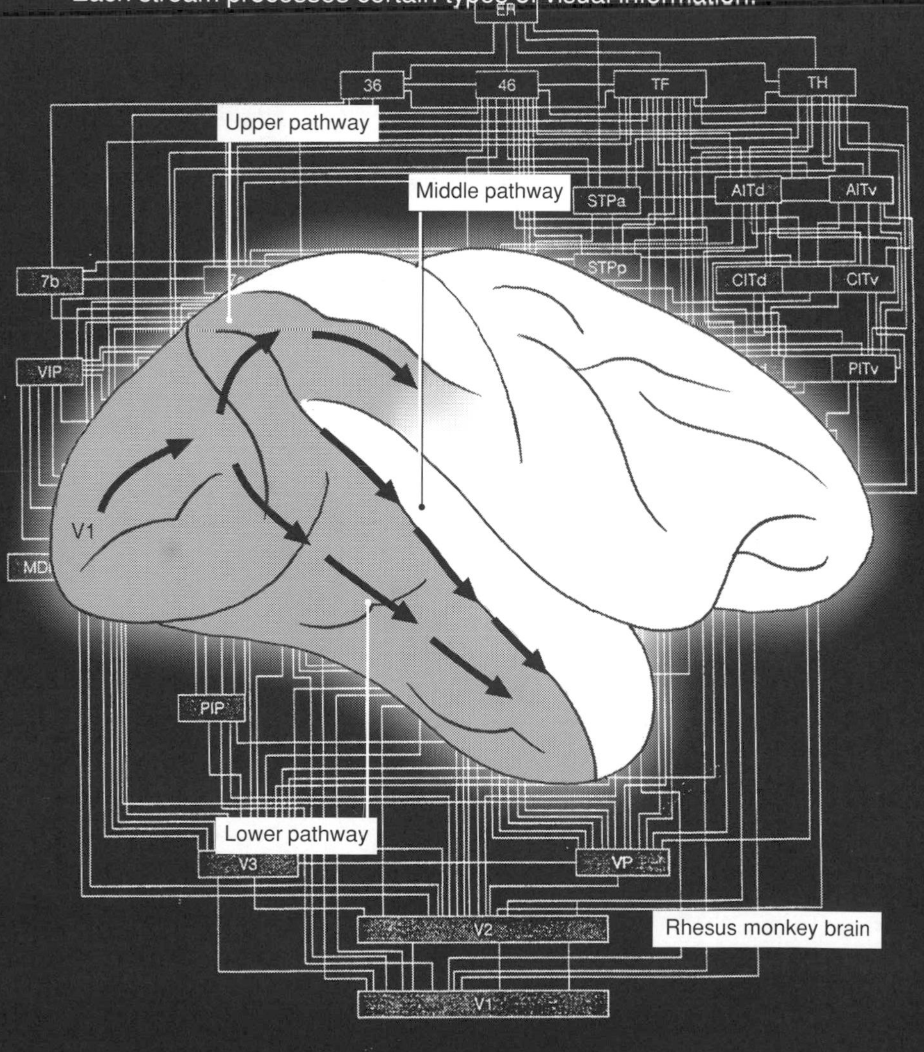

The Lower Visual Pathway: Effects of Injury on Recognition

Cells in the **temporal lobe** are choosy about what they respond to. Many of them raise their firing rate to faces, even to particular faces. Others like particular objects, for example, hands. These findings from electrode recordings in monkeys receive strong support from studies of people whose visual recognition has been affected by temporal lobe injury.

The inability to recognize objects is called **object agnosia**. There are several types. In **form agnosia**, the person sees colour, depth and contour but perceives only parts, not whole objects.

It is as though their attention skips from one fragment of contour to another without managing to piece the bits together.

Such people cannot draw a copy of what is in front of them, yet may be able to draw the same shape from memory.

In **simultagnosia**, objects are perceived and recognized, but only one at a time. The person cannot put together the various objects in a scene to understand it. When two objects he can recognize in isolation are presented in overlap, he has difficulty in visually separating them for recognition.

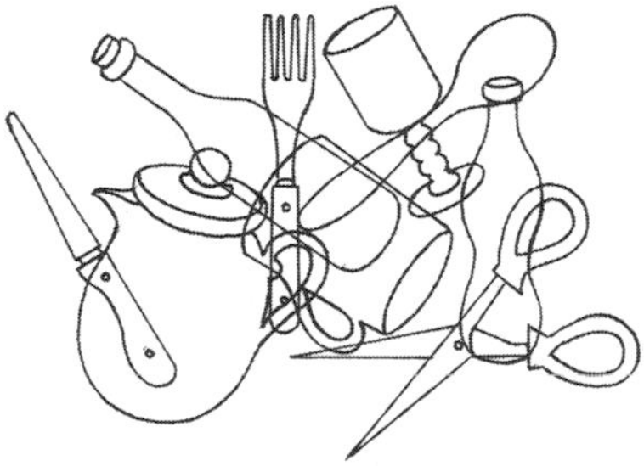

In **associative agnosia**, patients accurately describe or draw visual scenes and objects but there is a failure of **recognition**. The person cannot give the name or use of a glove or a fork. She may know the superordinate category to which an object belongs (clothing or cutlery) without knowing what it is (glove or fork). Despite this, she may know whether an object is real or imaginary.

In **prosopagnosia**, the problem is in recognizing familiar faces, often including the person's own. A prosopagnosic can still recognize voices. He can also describe a seen face, and even "read" its emotional expression, but he cannot make the leap to identity from the face alone. It seems that the **lower processing stream** has become **disconnected** from an emotional sense of familiarity generated in the limbic system.

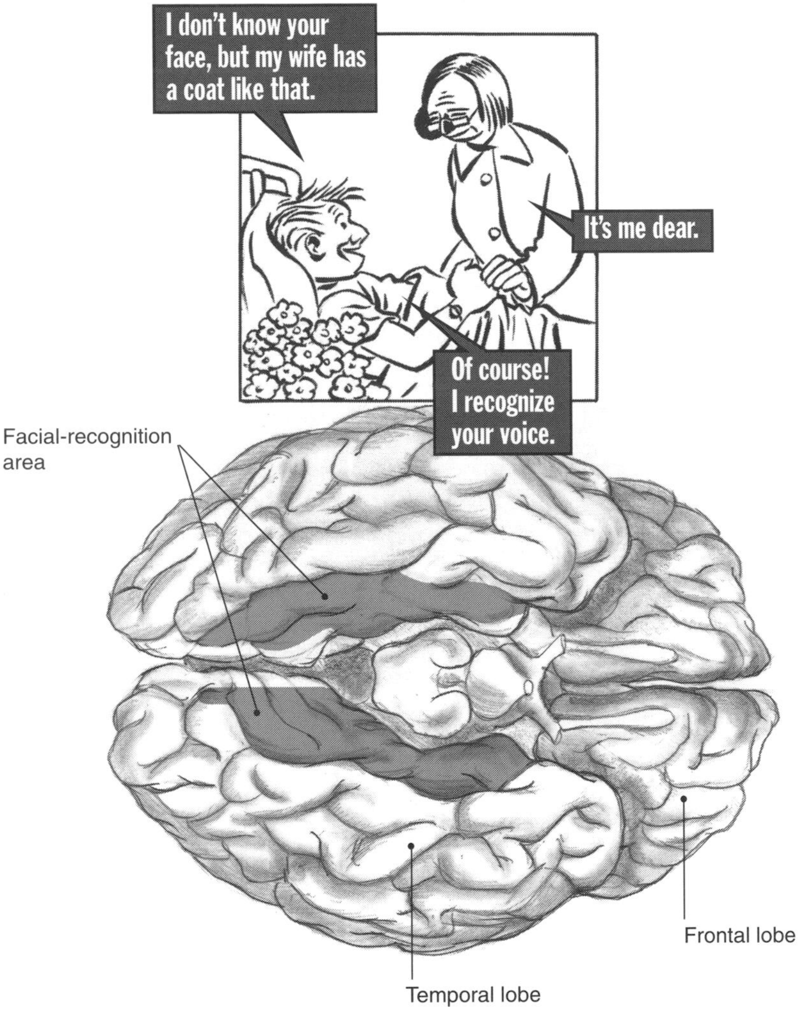

Although prosopagnosics do not **consciously recognize** familiar faces, they show the normal increase in bodily emotion to them (an increase in sweatiness!).

Also, when asked to learn pairings of faces with famous names (which they do recognize), they learn true pairings faster than false ones.

Albert Einstein

Diego Maradona

These results show that prosopagnosics still have both **emotional** recognition and **identity** recognition, probably in their upper visual pathway. These two kinds of recognition have, however, become disconnected from conscious visual experience. Occasional failure of this connection may underlie the not uncommon experiences of **déjà vu** (familiarity without recognition) and **jamais vu** (recognition without familiarity), both of which are frequent during episodes of **temporal lobe epilepsy**.

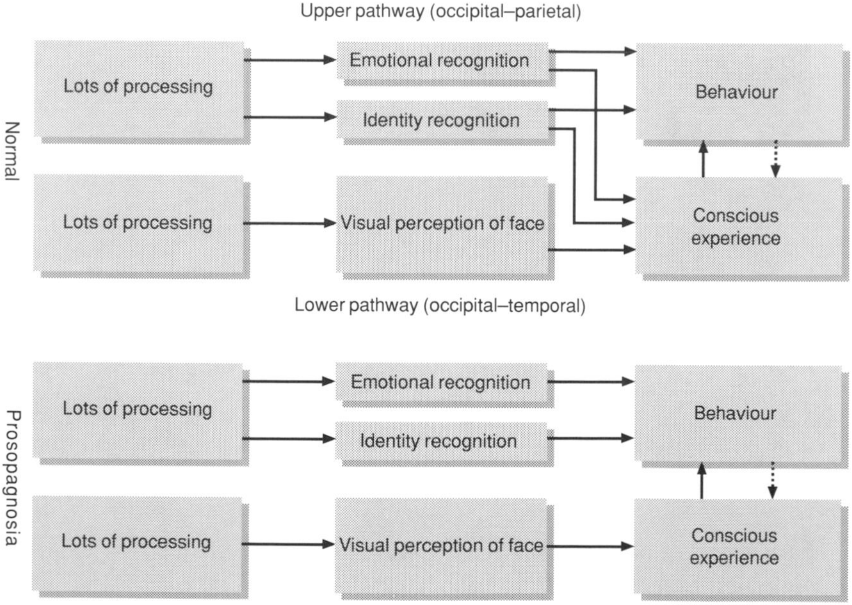

A Recognition Test

Prosopagnosia is especially likely after right temporal lobe damage. The following illustration allows you to experience for yourself the special role of the RH in face recognition: the "split faces" test.

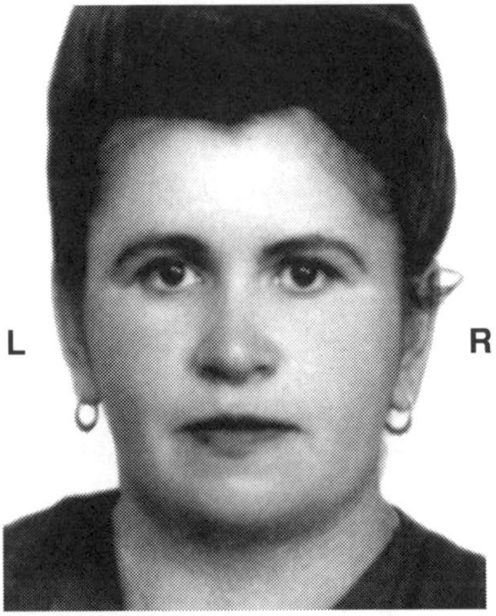

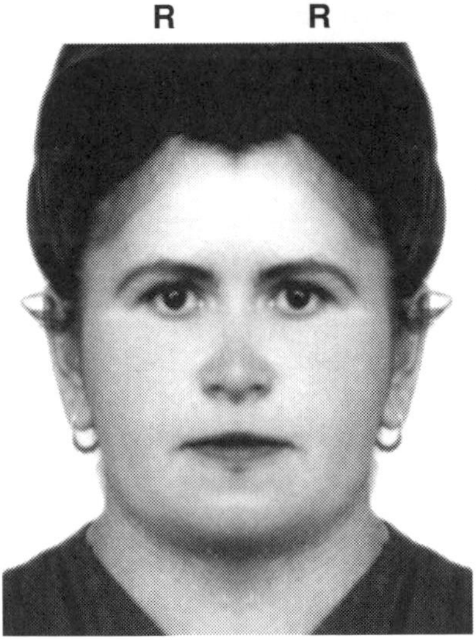

"I'm made from two right halves of the above face."

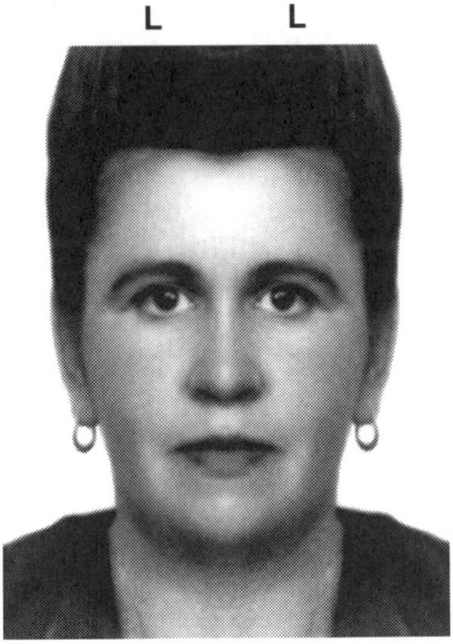

"I'm made from two left halves of the above face – people say I look more like her."

The left half of each face is seen by your RH (see pp. 106-7), which plays a greater role in face recognition than your LH.

The Middle Visual Pathway: Relative Spatial Positions

The middle visual pathway from the occipital lobe to the upper temporal sulcus is a recent discovery. It is poorly understood but may play a part in perceiving the **relative spatial positions** of objects. **Simultagnosia** could involve damage to this pathway, since if you can only see one object at a time then it follows that relative positions will be impossible to judge. Support for the idea comes from the finding that many simultagnosics have "route finding" difficulties in familiar environments.

WE OFTEN FIND OUR WAY BETTER IF WE CLOSE OUR EYES AND WORK FROM MEMORY.

The Upper Visual Pathway: Effects of Parietal Damage

Studies with monkeys show that many cells in the rear parietal lobe only fire during **reaches** towards an object. These cells may code information needed to **act upon** objects rather than to **perceive** them. For example, to pick up a book you need to "know" (not necessarily consciously) its location in relation to yourself, and its size, shape and probable weight.

In **Bálint's** syndrome, people with parietal damage can recognize objects accurately (using their lower pathway), but cannot accurately reach for them. These patients often fail to set their thumb and forefinger an appropriate distance apart when attempting to pick up an object.

They also fail to rotate their wrists to the correct angle when asked to "post" their hand into a slot, even though they can accurately report how the slot is tilted.

The lower pathway is responsible for **conscious visual perception**. The upper processing stream is responsible for **visually guided action** on objects, which is largely unconscious. The two processing streams certainly connect to each other, probably via the limbic and rhinal cortices. But further dramatic evidence that they can operate independently comes from an individual with **form agnosia**.

This woman can see flashes of light and makes very fine discriminations of colour. She can readily identify wooden letters by touch but is completely unable to do so by sight. Despite this, she does not bump into things and she can catch balls and sticks thrown to her. She can reach towards objects and, in grasping them, sets her grip to an appropriate size.

When shown a slot and asked to indicate its orientation by tilting a "postcard" in her hand to the same angle, she fails miserably.
Yet, paradoxically, when asked simply to post the card into the slot, she does so quite normally, rotating her wrist to the proper angle.
JUST DO IT!

This demonstrates that the upper stream has independent control of **unreflective** actions. However, when an action must be used to report what she consciously sees, then co-operation between intact streams is essential.

This section has presented a tiny fraction of what we know about visual perception in relation to mind. It turns out that the visual system works in very surprising ways.

Mind Spaces

Damage to the **parietal lobes**, especially the **right** one, impairs performance on many tests of **spatial ability**. Some of the most dramatic evidence that the right half of the brain is specialized in spatial skills comes from people with **split-brains**. Members of this small group have all suffered from extreme epilepsy. Their fits start in one side of the brain and spread to the other side through the 200 million fibres of the **corpus callosum**.

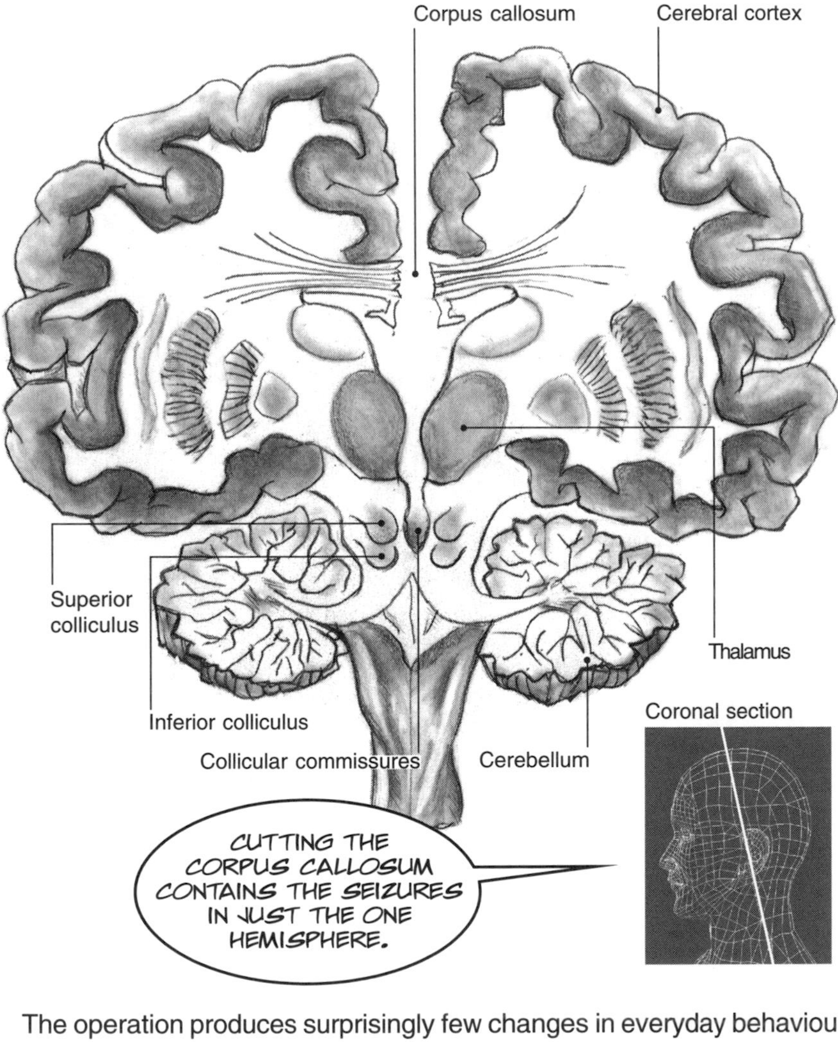

The operation produces surprisingly few changes in everyday behaviour and markedly reduces the frequency and severity of seizures.

One very curious finding is that after the operation previously right-handed people draw better with their **left** hands. (Performance with either hand is worse than pre-operatively.) This comes about because the left hand is controlled by the RH and the right hand by the LH. In intact brains, the two hemispheres share their abilities and knowledge via the corpus callosum, so both would contribute to the movements of a right hand.

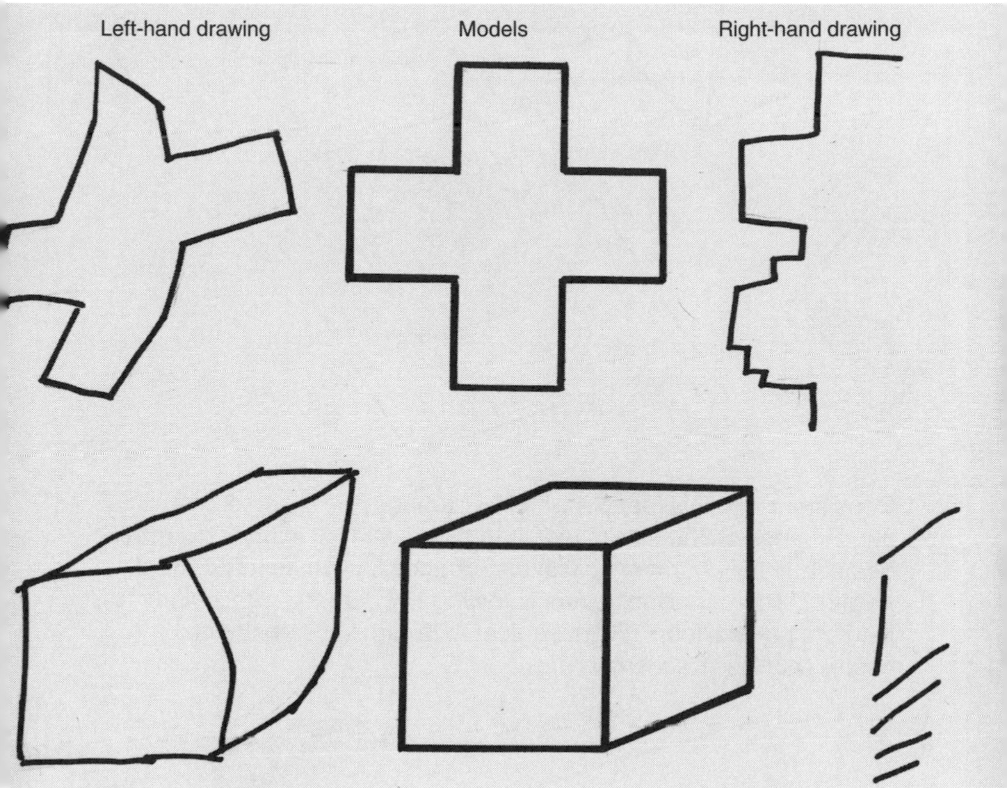

Left-hand drawing
Models
Right-hand drawing

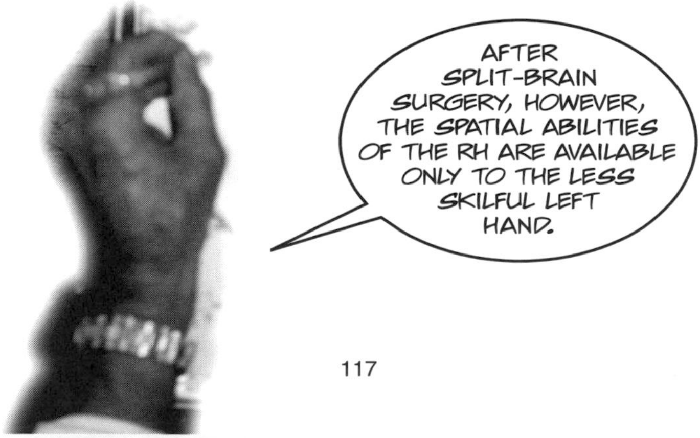

AFTER SPLIT-BRAIN SURGERY, HOWEVER, THE SPATIAL ABILITIES OF THE RH ARE AVAILABLE ONLY TO THE LESS SKILFUL LEFT HAND.

The superior spatial ability of the RH also shows up in a test in which coloured blocks must be arranged into a specified pattern. Split-brain people perform faster and more accurately with the left than with the right hand.

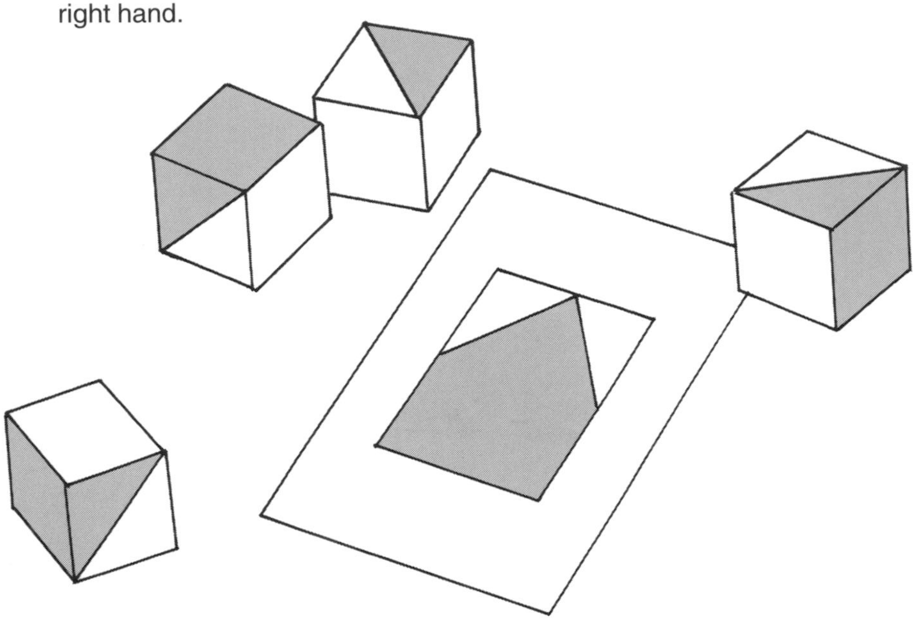

Consistent with this, people with RH damage show worse performance on the blocks test than those with LH injury. This may be because of a form of spatial disturbance known as **left spatial neglect**. This condition occurs following RH damage, particularly to the right parietal lobe. (Right neglect following LH damage also occurs, but less frequently.)

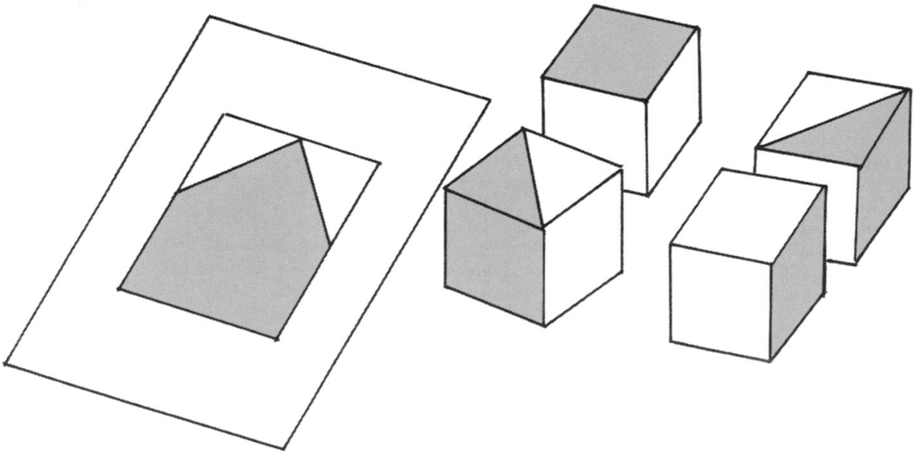

A person with left neglect may fail to dress the left side of his body or to eat the food on the left side of his plate. In bed, he may constantly turn to the right, falling out unless held in by side-bars.

In a standard diagnostic test, the person has to cancel all the lines on a page. People with neglect miss many lines on the left.

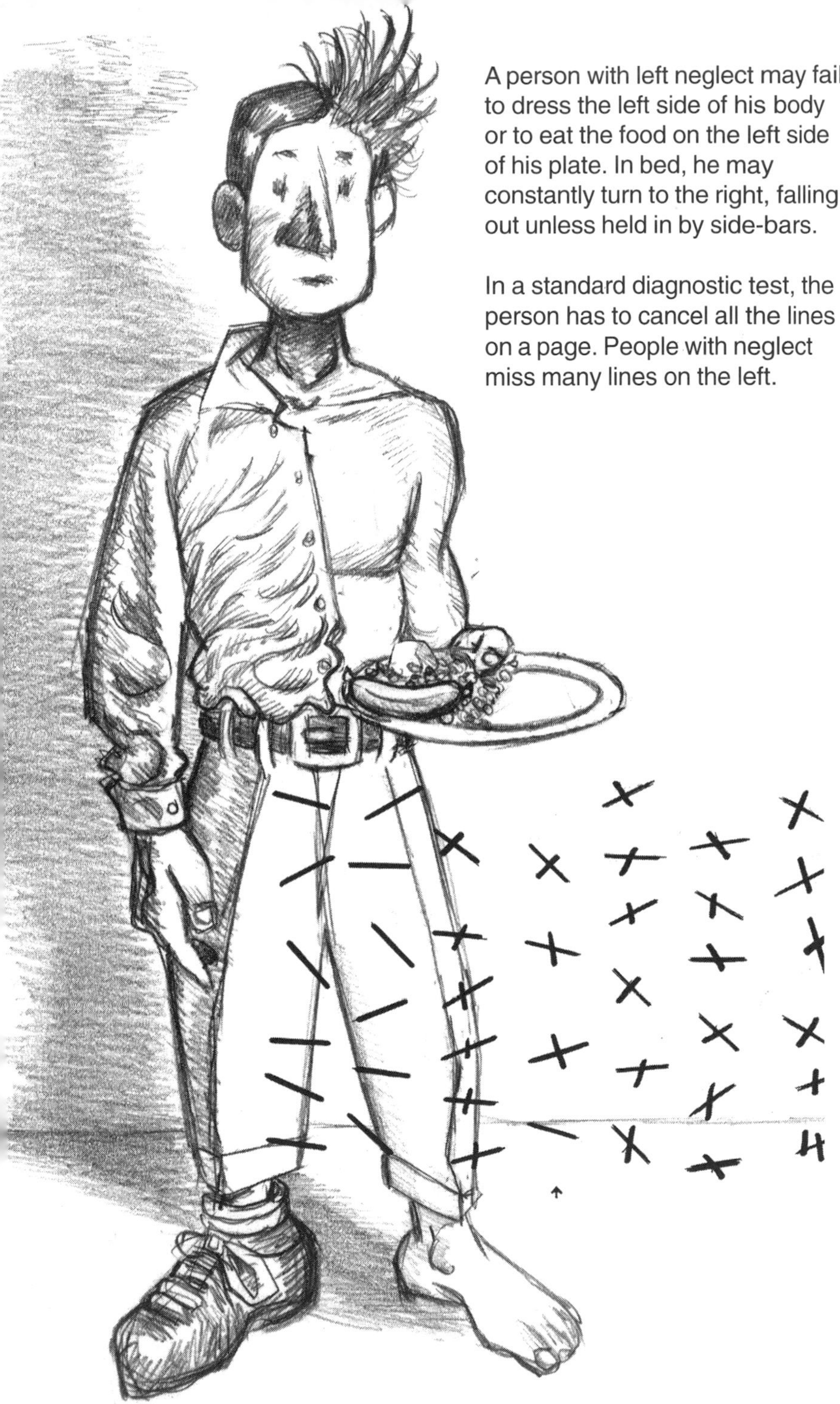

Visual, Motor and Imaginal Spaces

People with neglect are **not** blind to the left side of space; they can report the identity of a letter flashed in the left field. Generally, however, they ignore left space. Is this because they have difficulty attending to the left or because they cannot easily make motor movements to the left? Line cancellation requires them to do *both*, and experiments show they have *both* problems. Neglect can apply to both **visual space** and **motor space**. This seems complicated enough; but things get worse!

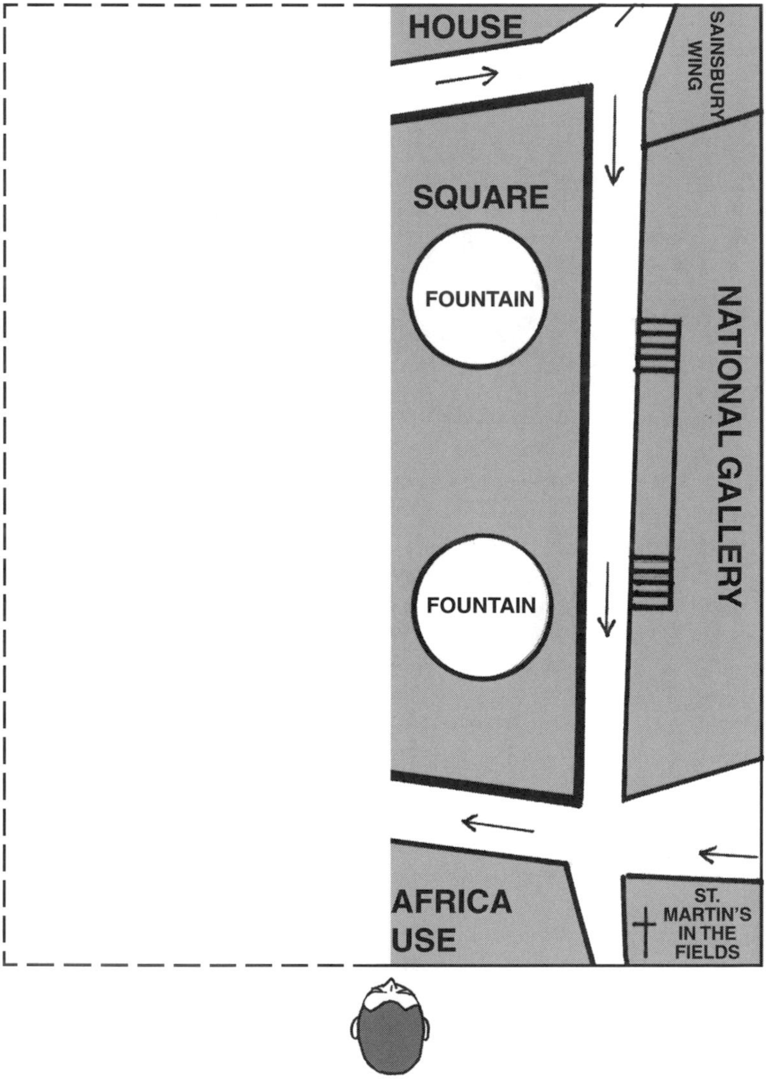

Suppose someone with neglect is asked to describe from memory or to draw, say, Trafalgar Square as seen from one side. His description omits all mention of the left side of the Square. If he then describes the Square seen from the opposite side, he now includes all the previously omitted details but leaves out all the previously included ones. So neglect applies not only to perceptual and motor spaces but also to **imaginal space**.

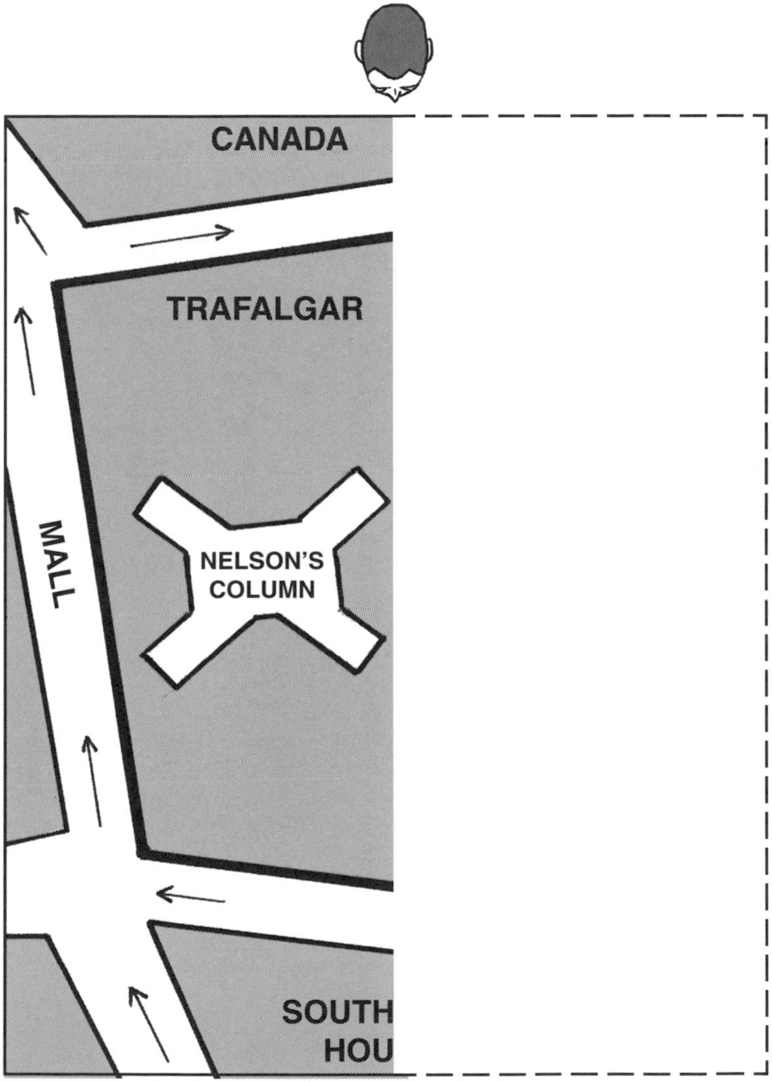

Representations of Space

It seems that the RH, and especially the right parietal lobe, somehow specializes in constructing **representations** of space. Tests that require a person with left spatial neglect to set up different kinds of representation can all show left neglect.

People use many kinds of (usually unconscious) spatial representations.

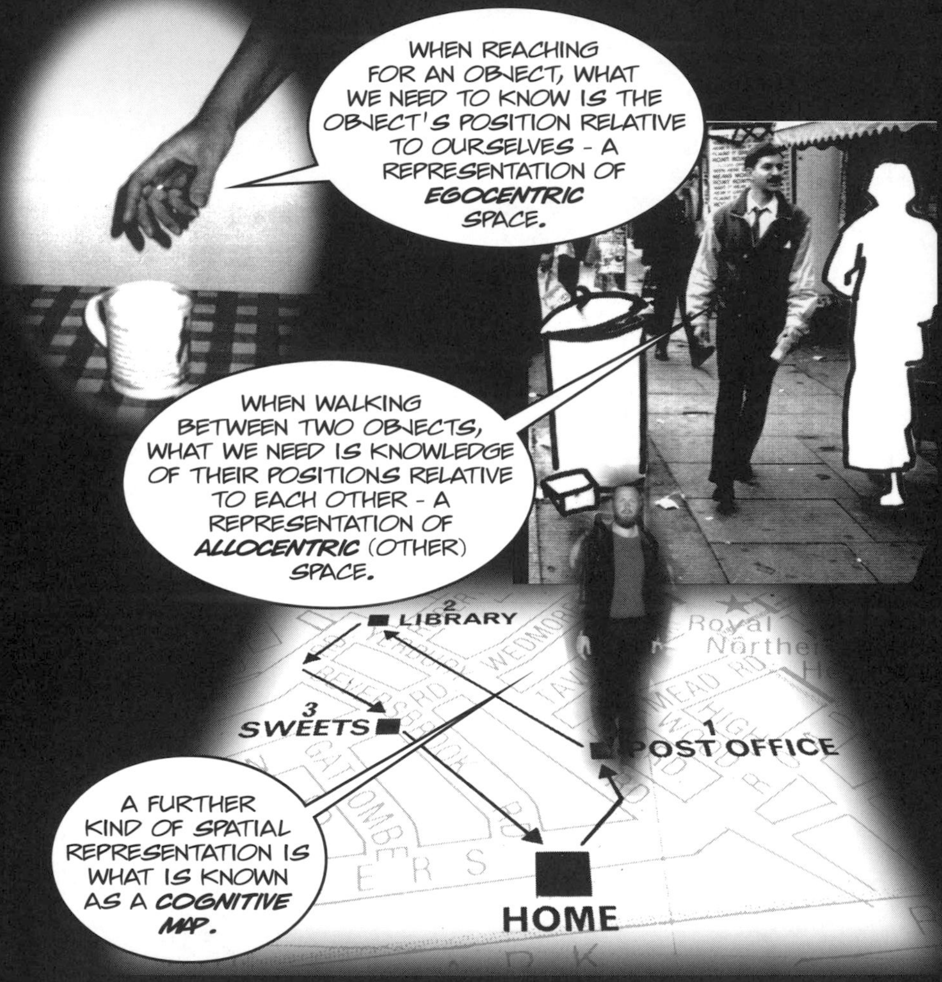
WHEN REACHING FOR AN OBJECT, WHAT WE NEED TO KNOW IS THE OBJECT'S POSITION RELATIVE TO OURSELVES - A REPRESENTATION OF ***EGOCENTRIC*** SPACE.
WHEN WALKING BETWEEN TWO OBJECTS, WHAT WE NEED IS KNOWLEDGE OF THEIR POSITIONS RELATIVE TO EACH OTHER - A REPRESENTATION OF ***ALLOCENTRIC*** (OTHER) SPACE.
2 LIBRARY
3 SWEETS
1 POST OFFICE
HOME
A FURTHER KIND OF SPATIAL REPRESENTATION IS WHAT IS KNOWN AS A ***COGNITIVE MAP***.

This refers to the layout of places and objects, and of routes between them. Cognitive maps include details of locations that are not presently observable; yet many animals, including rats, possess them.

Cognitive maps are closely associated with a limbic structure, the **hippocampus**.

The hippocampus takes its name from a supposed resemblance to a mythological sea-horse.

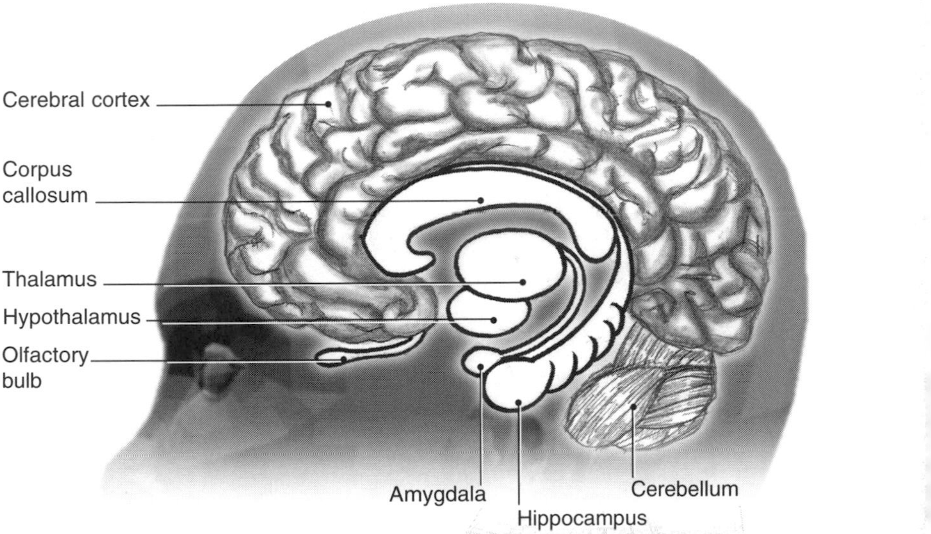

People who suffer damage to the hippocampus become bad at finding their way about. Some, if they remain in their own home, can cope with the familiar environment. However, a change of address, perhaps to residential care, leaves them forever disorientated.

Others lose even their long-established cognitive maps and have difficulty getting from room to room in their own house.

Clearly, there is a lot to learn about how the mind and brain deal with space.

Attention and the Mind

If the mind carries out actions in mental space(s), rather as the body does in physical space, then modern studies of **attention** illustrate nice parallels between these inner and outer worlds.

THE MOST BASIC ACT OF ATTENTION IS THE ORIENTING RESPONSE, A TURNING OF THE BODY TOWARDS SOME NEW OBJECT OR EVENT IN ORDER TO FIND OUT MORE ABOUT IT.

In some animals, whole body movements are replaced by orienting of just the **sensory apparatus**. Dogs cock their ears towards a sound source and many animals move their eyes to fixate (focus on) changes in the environment.

In humans, and at least some other primates, attention can become a purely *mental* act. We are capable of attending away from our fixation point.

This may be the origin of our ability to deceive, and also to dwell in the imagination on selected memories or possible futures.

Experiments with Attention

Cueing experiments demonstrate the separation of attention and fixation. Suppose you stare at a central square on a monitor screen. Either a directional cue (< or >) or a neutral cue (+) appears briefly in the square.

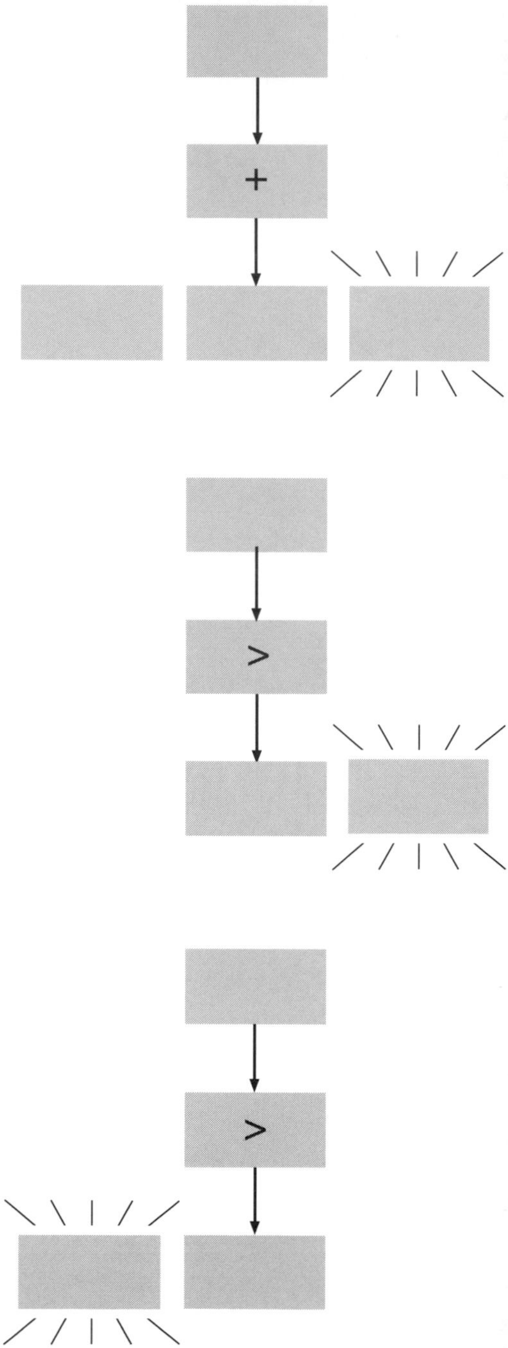

Then a target square flashes either to left or right of the centre square, and you have to press a response button as fast as possible.

Reaction time is faster when the directional cue points in the actual direction of the subsequent target (valid cue) than when there is only a neutral cue.

In other words, the cue shifts attention towards **where the target will subsequently appear**, and this shows up as a faster response. Conversely, if the cue points in the wrong direction (invalid cue), reaction time is slower than on neutral trials. These events occur too fast for eye movements to be made. The effects depend on movement of an *internal focus of attention.*

The Attentional Network

A network of brain areas (**parietal lobes**, **pulvinar**, **superior colliculi**) seems to mediate spatial attention. Brain imaging reveals increased activity in the parietal lobes during spatial shifts of attention, and damage to the back of these lobes impairs shifting.

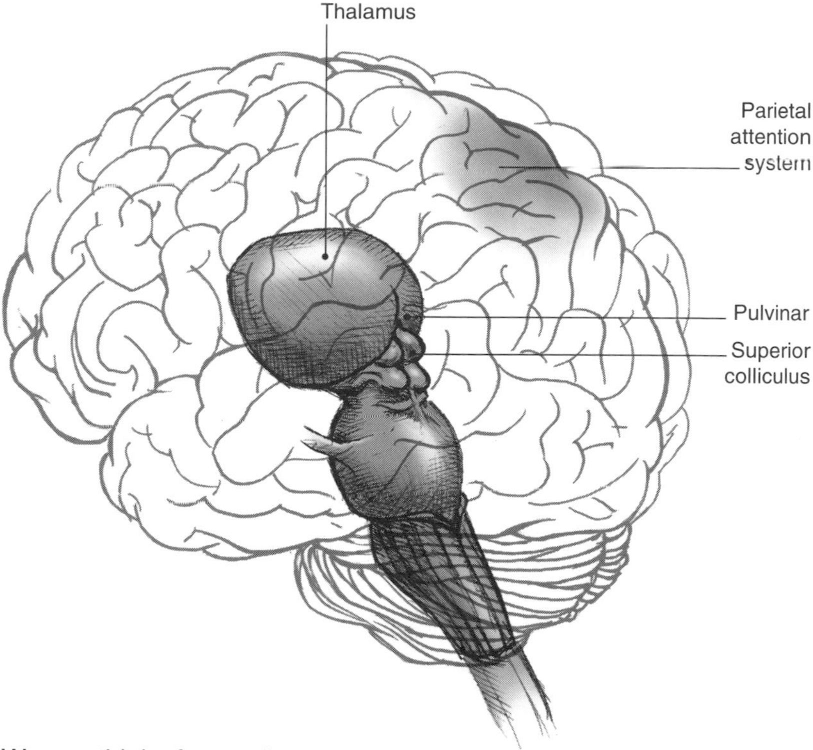

We can think of *attending* to an object as the mental equivalent of *picking it up*. So far we have been thinking of just the **reaching**, or spatial, component. There is also a grasping component to consider. When you reach for an object, you find that your hand arrives already shaped for what it will grasp. The preparatory shaping is controlled, unconsciously, in the upper visual pathway.

Mental Grasp

In visual attention, too, the mind “grasps” an object for which unconscious processes have prepared it. When you look at the figures below, you do not see a mass of unconnected lines and patches. You see individual three-dimensional shapes.

The reaching and grasping components of attention are known as **space-based** and **object-based** attention. We can see the difference by asking a person with left spatial neglect to cancel lines that form two separate blocks on a sheet of paper.

With a single block, the person neglects all the lines in left space. With the separated blocks, however, he cancels some of the lines on the **right side** of the **left block**. Similarly, he would normally cancel all the lines in attended right space. With separated blocks, however, he ignores some lines on the **left side** of the **right block**.

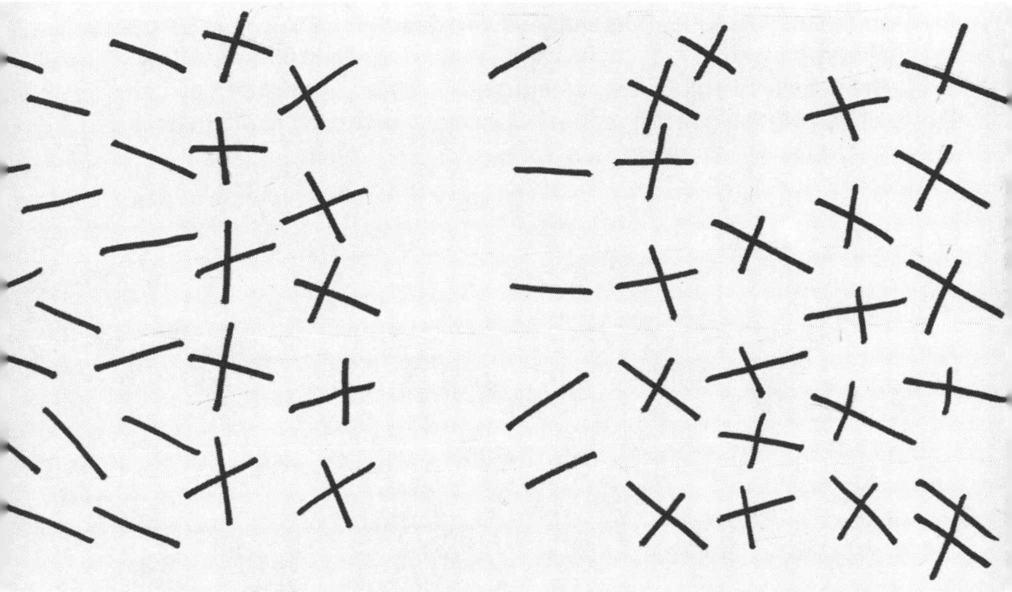

The person shows two sorts of left neglect. Neglect of the left side of space involves space-based attention. Neglect of the left side of objects involves object-based attention. (In this example, a **block** of lines is a perceptual object.) Both kinds of neglect apply to the left block, so the majority of its lines are ignored. Only object-based neglect applies to the right block, so the majority of its lines are cancelled.

Currently, it is thought that damage to the upper processing stream (occipital→parietal) causes space-based neglect, while damage to the lower processing stream (occipital→temporal) results in object-based neglect.

What is Consciousness?

The word "consciousness" has a variety of meanings. Consider that when we sleep we are **unconscious**, and yet in dream sleep our visual and emotional experiences are vividly **conscious**. The first sense of "consciousness" refers to a state of wakefulness or arousal. The second sense identifies "consciousness" with sensory and emotional experience.

Various brainstem structures control consciousness in the sense of wakefulness. They include the **reticular formation**, the **pons**, the **raphe nuclei** and the **locus coeruleus**. Stimulation of the reticular formation increases wakefulness and destruction of it induces **coma**. Contrastingly, lesions of the raphe nuclei lead to **insomnia**. However, the activity of both these structures is normally modulated by the locus coeruleus and the pons. Consciousness-as-wakefulness is controlled by a network of centres.

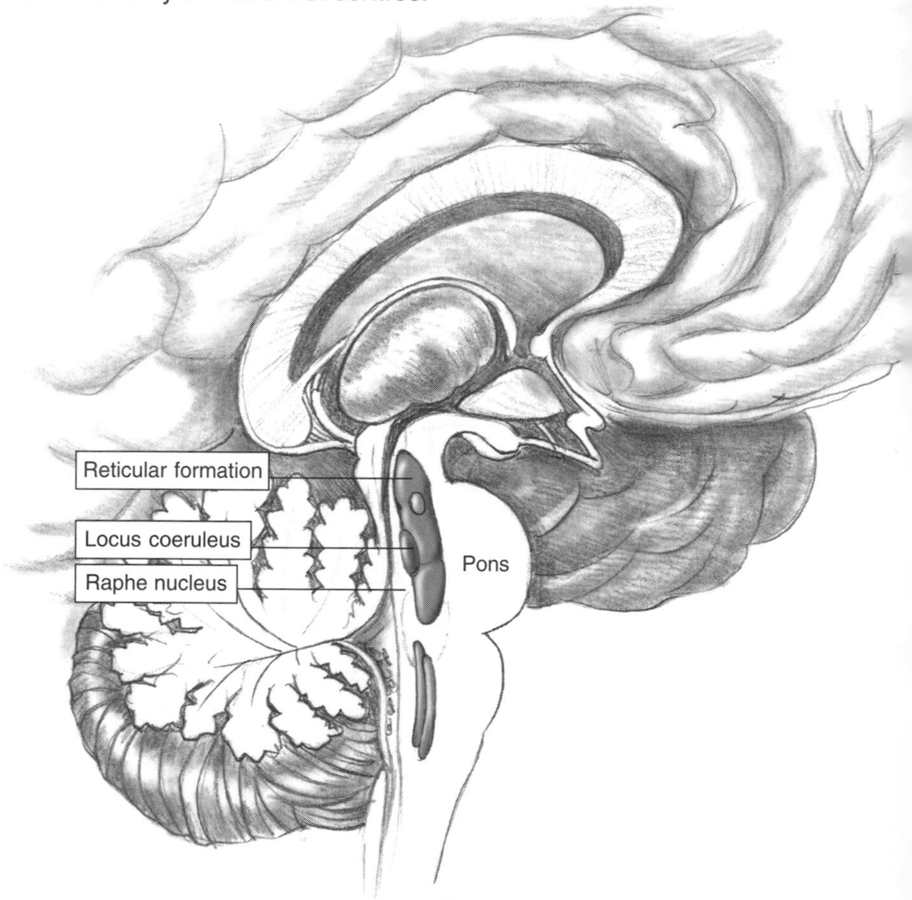

Consciousness understood as **sensory experience** poses many puzzles. Damage to a restricted region of visual area V1 produces an island of blindness in the visual field, a **scotoma.** If a light is flashed in a person's scotoma she does not report it, though she reports lights just outside the scotoma as normal. Somebody with a **scotoma** is as unconscious of it as we all are of the blind spots in our eyes.

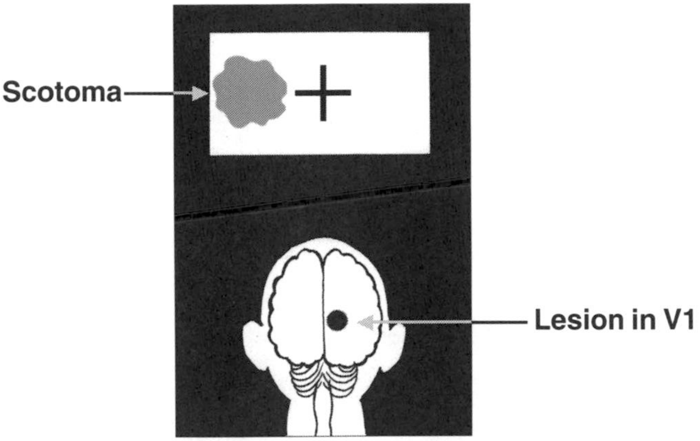

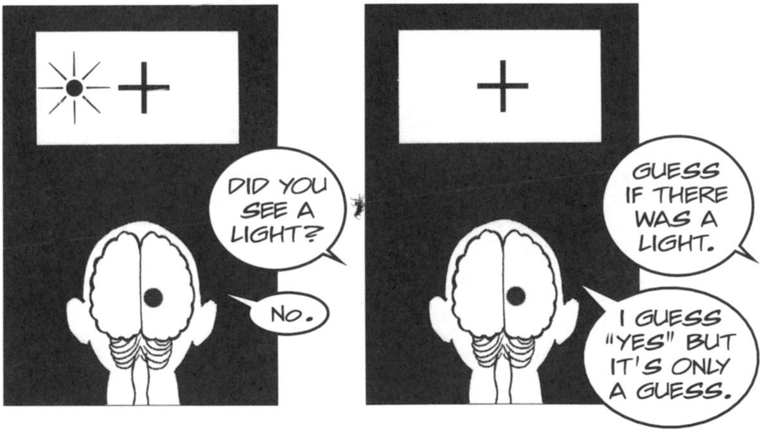

Curiously, however, although they have no conscious experience of lights flashed in a scotoma, people may be able to report accurately whether or not a light was flashed on a particular trial. When told this, they are incredulous, and have to be persuaded of their ability.

Yet they can also discriminate horizontal from vertical lines, and stationary from moving targets, all the while convinced that they are guessing. This phenomenon is **Blindsight**.

Blindsight

Blindsight is in part due to a sparse set of fibres that run directly from the lateral geniculate nucleus to visual areas V4 and V5, bypassing V1. The purpose of these fibres is unknown. What is certain is that, while conscious visual experience requires an intact V1, some visually-controlled behaviours do not require consciousness.

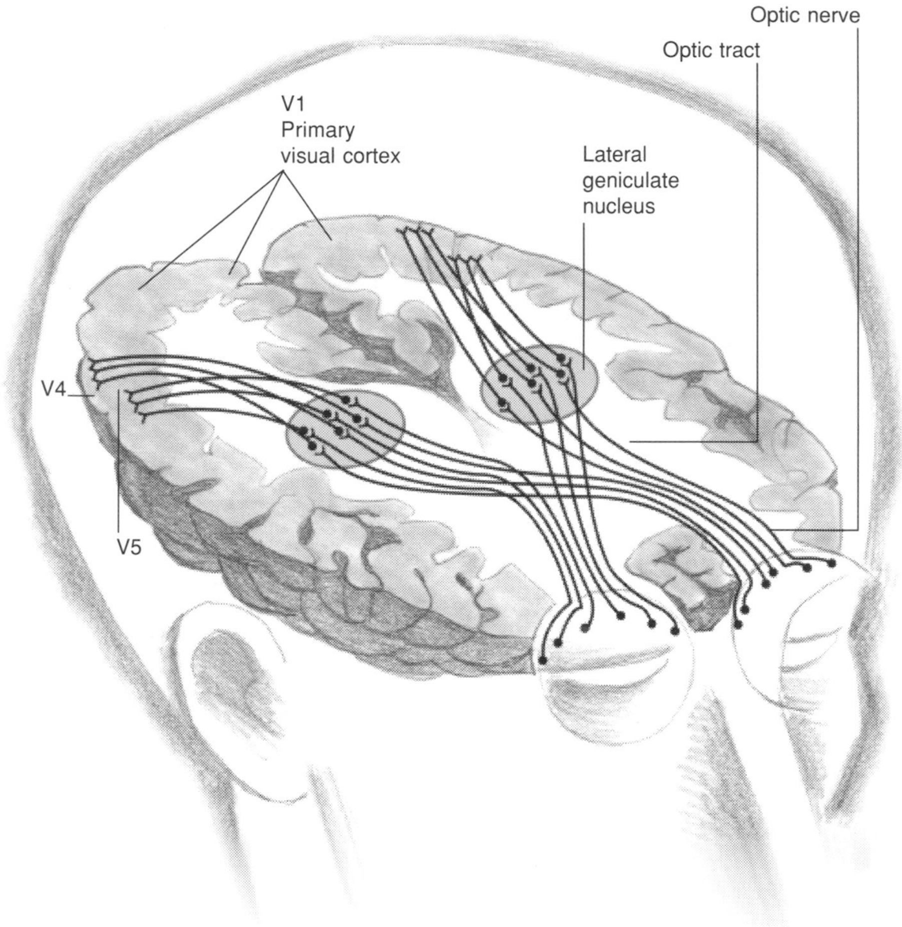

Consciousness raising takes place in political and psychotherapeutic groups, the members of which might be suddenly afflicted with **self-consciousness** when called upon to speak. In these two uses, "consciousness" seems to refer to the contents of our thoughts. Consciousness is raised when we become aware of oppressions suffered or inflicted. Self-consciousness occurs when the focus of awareness shifts from others to ourselves.

Working Memory

Consciousness in the sense of the contents of our thoughts, what we currently "have in mind", is frequently studied under the heading of **working memory**.

3.658
1.416
6.729
11.803
11.803
2 608
307
14.718
213
6.028
126
6.367
21.085
WORKING MEMORY IS WHAT YOU USE TO ADD UP A BILL IN YOUR HEAD AND KEEP TRACK OF THE SUB-TOTALS.
OR TO REMEMBER WHERE YOU ARE IN A SENTENCE OR AN ARGUMENT.
OR TO SWITCH BACK AND FORTH BETWEEN A GAME OF CHESS AND PREPARING A MEAL - SO LONG AS YOU DO NOT CONCENTRATE FOR TOO LONG ON EITHER TASK.

Working memory briefly stores and processes information needed in planning and carrying out tasks. It has three parts. Most important is the **central executive**, or decision maker, with other systems slaved to it.

A **visuo-spatial system** represents limited information about spatial relationships.

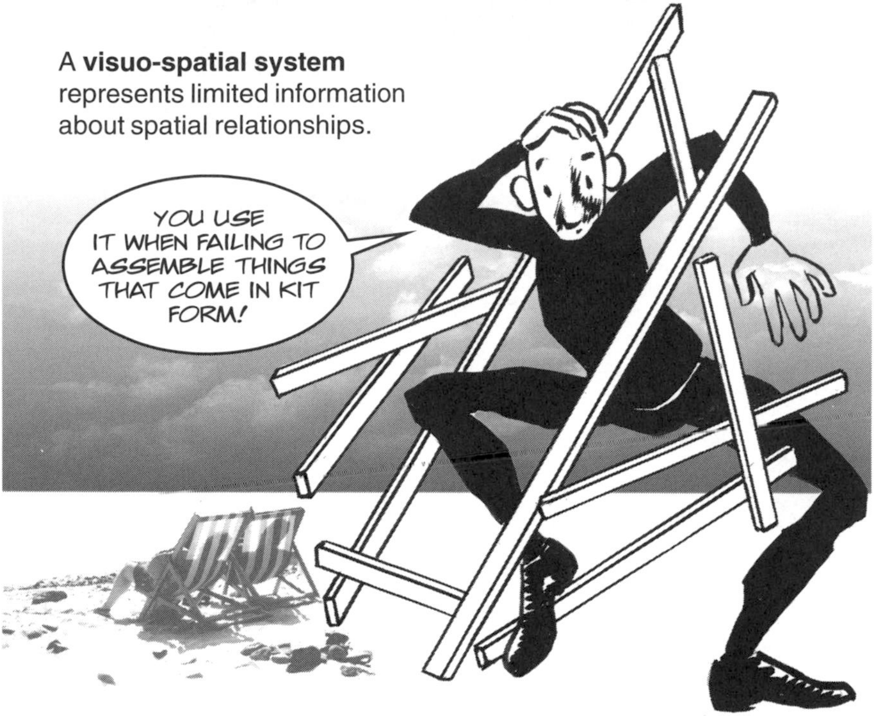

YOU USE IT WHEN FAILING TO ASSEMBLE THINGS THAT COME IN KIT FORM!

An **audio system** allows you to hold on to a limited number of words while you re-arrange them into more intelligible phrases, or work out their meaning.

In recent years, brain images, lesion studies and electrode recordings have shown that:

- various regions of the LH contribute to *verbal working memory tasks*
- various regions of the RH are involved in *spatial working memory tasks*

In all cases, there is also activity in the frontal cortex.

The Central Executive in Area 46

Although different tasks seem to recruit varying regions of frontal cortex, one area does seem common to them all. **Area 46**, as it is known, is currently the favourite candidate for the role of **central executive**.

Working memory draws on areas throughout the cortex.

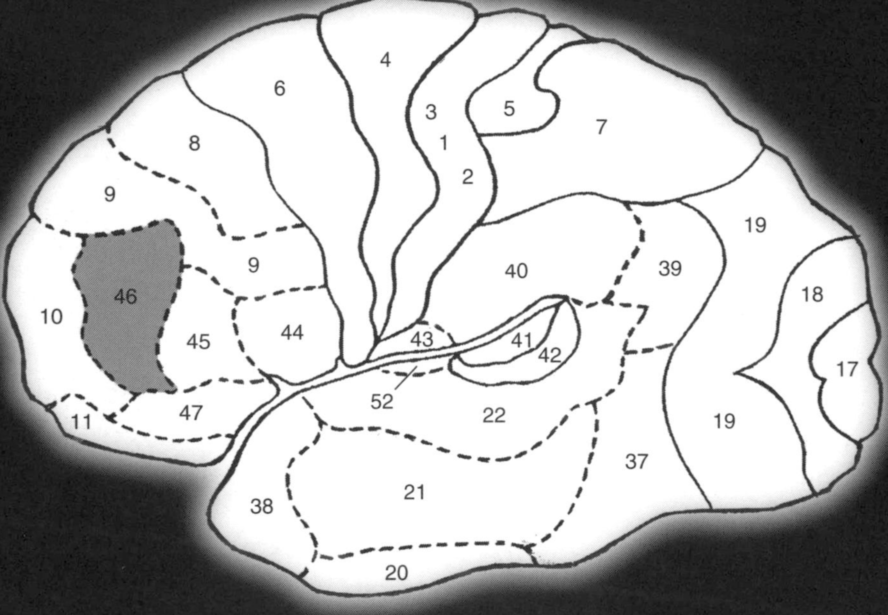

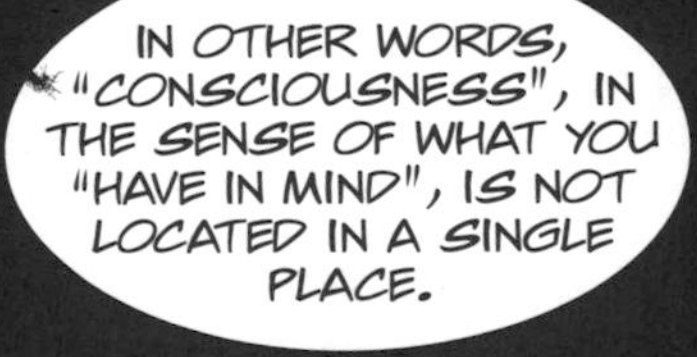

Area 46 may be vital to co-ordinate your thoughts and to switch back and forth between tasks. However, the **content** of consciousness depends on which areas of which hemisphere are momentarily engaged by the task in hand.

Because the frontal cortex of *each* hemisphere has its own area 46, a **split brain** individual can (or can appear to) possess double consciousness.

Suppose two pictures are simultaneously flashed, one to each hemisphere. If the split brain person is asked to **say** what she **saw**, the verbal LH reports "apple". The visual areas, verbal areas and area 46 of her LH work in concert to give this response. However, if she is asked to **write with her left hand** what she **saw**, she writes "spoon". Now co-operation is between the visual areas, motor control areas and area 46 of the RH.

If the split brain person is then asked to explain her two responses, her speaking LH has a problem. It/she does not know why the RH caused the left hand to write "spoon". To avoid embarrassment, it/she **confabulates** an explanation, that is, devises imaginary experiences.

This is an example of **narrative consciousness**, the constantly rehearsed and revised story of our self that each of us tells.

Free Will and the Frontal Lobes

When **Penfield** stimulated the motor cortex of conscious surgical patients, they assured him that they experienced their consequent movements as involuntary, not willed.

The **motor cortices** are at the back of the frontal lobes (FLs). Their role is to initiate execution of cortically generated movements, as opposed to movements generated sub-cortically or spinally (as we saw in the section on movement). But Penfield's patients bear eloquent witness that they are not **the seat of the will**.

Responsive Movements

In front of the motor cortex are the **premotor** cortex and **supplementary** cortex. These areas **select** movements that the motor cortex will execute.

Premotor cortex selects movements in response to *external* triggers.

AS WHEN YOU RISE FROM YOUR SEAT IN RESPONSE TO THE TELEPHONE RINGING, OR KEY IN A TELEPHONE NUMBER READ FROM THE DIRECTORY.
AS WHEN YOU RISE FROM YOUR SEAT IN RESPONSE TO FEELINGS OF DISCOMFORT, OR KEY IN A TELEPHONE NUMBER FROM MEMORY.

Supplementary cortex selects movements in response to *internal* triggers.

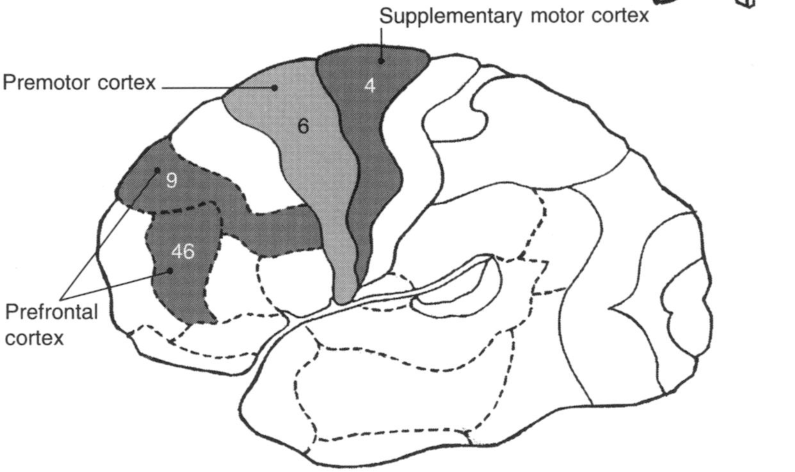

Forward of premotor and supplementary cortex lies the **prefrontal** cortex. This area has many incoming and outgoing connections. The upper and lower visual pathways from the parietal and temporal lobes terminate here.

Effects of Frontal Lobe Damage

It is difficult to characterize the role of prefrontal cortex, which includes area 46. Its functions include **sequencing** of behaviour and memory for **temporal** order. When pre-frontally damaged people have to copy a series of movements, they tend to reproduce the correct movements in the wrong order.

They also exhibit **perseveration** (excessive repetition), or behavioural **rigidity**. An example is performance on The Uses of Objects Test in which you have to suggest different uses for a particular object.

People with FL damage find this test very difficult. They repeatedly give the most common use.

They fail to inhibit the most obvious answer so that less obvious ones can come to mind and be expressed.

FL Damage and Unwanted Responses

Failure to inhibit unwanted responses also shows up in **environmentally driven** behaviour. Individuals with frontal lobe (FL) damage often react in stereotyped ways to objects they encounter, however socially inappropriate the setting. Seeing a toothbrush, they may pick it up and use it, even though it belongs to someone else and they are not in a bathroom.

Entering someone's home, they may overtly inspect the pictures on the walls, commenting upon them and pricing them as though in a gallery.

When the inappropriateness of their behaviour is pointed out, they may become confused or **confabulate** fantastic explanations of their actions.

Because they are so much at the mercy of environmental triggers, FL damaged individuals have great difficulty in formulating plans and following them through. Trains of thought and action get side-tracked by irrelevant associations (a characteristic shared by schizophrenics). They also have memory problems, when remembering requires the use of strategy: for example, the response of a normal witness to a lawyer's question …

FL individuals may also lack spontaneity, and be emotionally indifferent to themselves and others. This can occur without any loss of intelligence. They may respond reasonably to factual or theoretical questions but never initiate conversation or volunteer information.

What is Free Will?

Primates, and particularly humans, have large FLs (frontal lobes). We have seen that FL functions include the making of plans and the inhibition of unwanted behaviours, but are the FLs the much-sought **seat of the will**?

William James (1842-1910) considered that the sense of free will comes from having both **a conscious image of a goal** and **a conscious desire to achieve it**. We might add to these **knowing how to achieve the goal**.

Knowing how to achieve a goal involves being able to plan and to follow through the plan, avoiding distractions. Clearly the FLs, and more especially the pre-frontal cortex, are crucial to these functions. The inertness of some FL patients suggests that the FLs may also be essential for conscious desires. However, the FLs play a much smaller part in conscious imagining of goals.

Visual images of goals arise in the occipital→temporal regions of the lower visual pathway. Motor images of what to do to achieve a goal arise in the parietal→frontal regions of the upper pathway.

We have also already encountered the idea that voluntary action is based on *self*-instruction. This involves language zones in the left temporal lobe as well as the left FLs.

Clearly, willed actions are assembled using many brain areas.

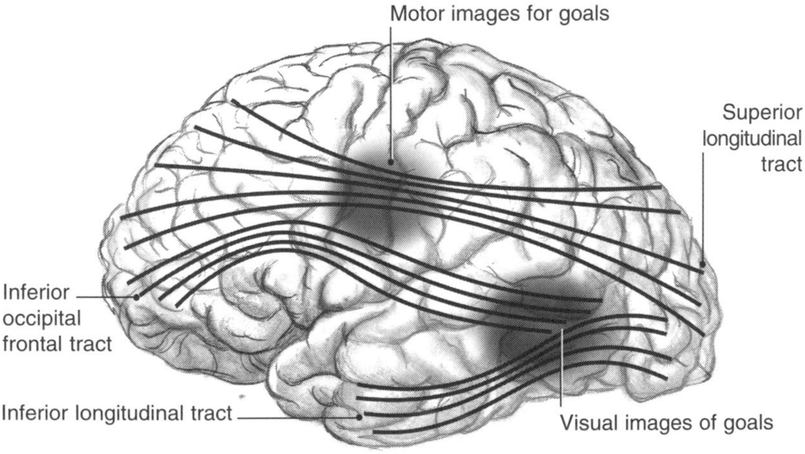

Perhaps, in considering free will, it is best to go back to Homer.

Odysseus, returning from Troy, longed to hear the Sirens, whose enchanting song lured sailors onto the rocks. He instructed his companions to tie him to the ship's mast and to fill their ears with wax. Temporarily deafened to both the enticements of the Sirens and the pleadings of their leader, the sailors held their course safely past the wreck-strewn shore where the Sirens perched.

Wily Odysseus recognized that the frontal lobes do not always have sufficient inhibitory control to match the power of compulsion. In doing so, he freed himself to experience the rapture of the Sirens' song.

The Self

Many strands go to make a sense of self.

The **social self** is the sum of the groups to which a person belongs.

The **inter-personal emotional self** is forged in relationships.

It says: he was a mighty hunter, an intrepid explorer, a heroic war leader, a statesman of world-wide renown and a disappointment to his mother.

Both these selves are beyond the scope of neuropsychological investigation.

However, we are on firmer ground, neuropsychologically speaking, with the cognitive or **narrative self**.

When the speaking left hemisphere of a split brain individual tries to explain behaviours controlled from both the LH and the RH, they provide a model of the circumstances in which we all find ourselves.

Each of us has to explain our behaviour, even though much of it may be mysterious to us. Our accounts are couched in terms of one of the accepted narratives of our culture.

They are anchored in three words with which everyone identifies: **my name**, **I** and **me**.

The narrative self haunts the linguistic regions of the LH and those many other cortical and sub-cortical areas that contribute to language. It is also crucially dependent upon episodic memory; and since autobiographical memories are located throughout the brain, the **narrative self** is necessarily widespread.

Loss of Self

Amnesics, of necessity, possess a damaged narrative self. Able to remember events from twenty years ago but not those of the past five minutes, the amnesic is stuck with the narrative self he had at the time of his injury or disease. Like the FL individual, his efforts to make sense of the anomalies and contradictions of his situation drive him to **confabulation**.

Here is an example of an amnesic in a hospital ward who believes he is still serving in his pharmacy.

Confabulations are attempts to maintain and update the narrative self.

The **bodily self** (or **proprioceptive self**: see section on movement) also lodges at diverse addresses throughout the brain. They include the sensory cortex, the thalamus and the cerebellum. This body self is largely unconscious. We only notice it when it goes missing. For most of us this means the strange effects of an injection at the dentist or a brief attack of "sleeping leg". Those individuals permanently deprived of proprioception suffer a devastating loss of self.

The loss cannot be easily verbalized but is strikingly illustrated by the joy one woman obtained from the feel of the wind against her skin. Although she had lost proprioception, she still had the skin senses of temperature, pain and, most importantly, touch.

Denial of Loss

Certain people undergo a partial loss of body self. This follows stroke or tumour damage to the right sensory cortex and its connections with mid-brain and frontal areas. **Anosognosic** individuals deny that they have left side paralysis and show no distress about it.

Even when they have repeatedly had to confront the fact of their defect, anosognosics never achieve more than a momentary acknowledgement of it. At best, they may admit previously having had problems with movement, but they deny any continuing difficulty.

Dissolution of Self

The **animal** self is the basic, biological sense of individuality. It distinguishes self from non-self. One effect of psychedelic drugs is to break down, or greatly weaken, this boundary. Knowing where in the brain the drugs act might help to pinpoint a location for the animal self.

One site of action is the **locus coeruleus** (LC), a cluster of neurons in the brainstem that funnel and integrate sensory inputs. Psychedelic substances alter activity in the LC. However, psychedelics act on a variety of structures, particularly **serotonin** pathways, so it is probable that even this core self is not identifiable with one particular area.

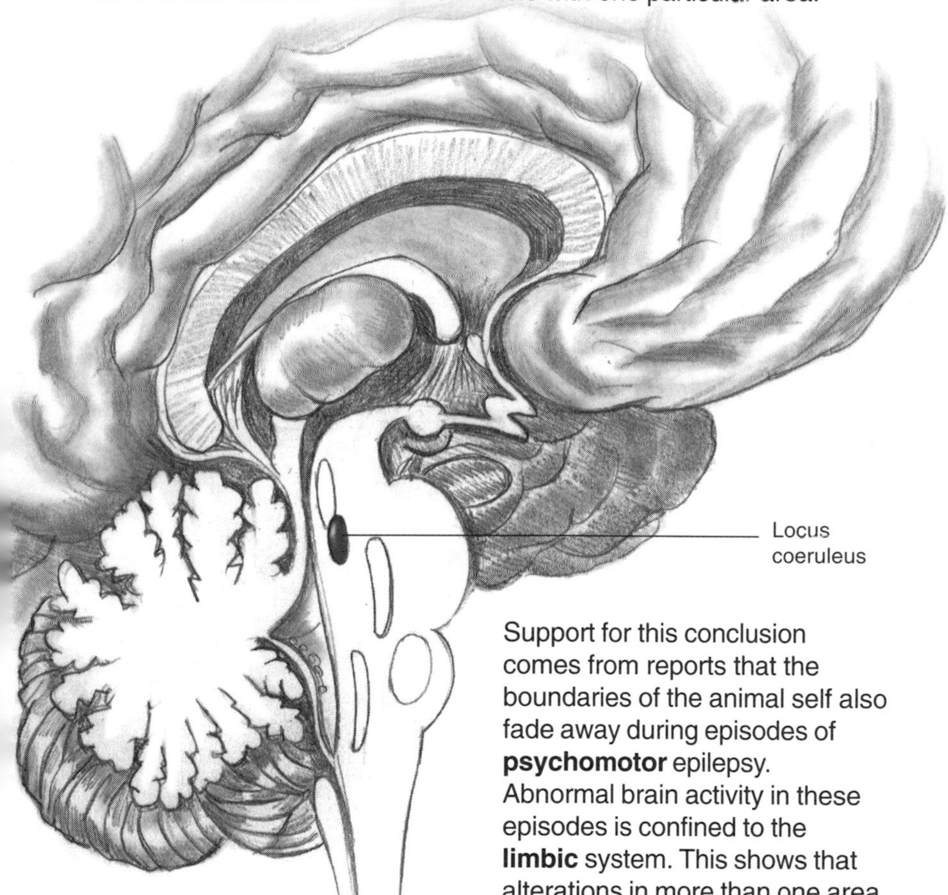

Support for this conclusion comes from reports that the boundaries of the animal self also fade away during episodes of **psychomotor** epilepsy. Abnormal brain activity in these episodes is confined to the **limbic** system. This shows that alterations in more than one area can produce a loss of animal self. Like our other selves, the animal self has no single address.

Feelings of Transcendence

Psychomotor epileptics and users of psychedelic drugs share more than just an experience of oneness with everything. Both are also prone to "gratulant" (as in con**gratulate**) feelings of fulfilment, triumph and elation. Both may experience a feeling of certainty, of "that's how it is and that's how it has to be". Yet, although such feelings are sensed with great conviction, they do not attach to any particular thing. They are free floating.

ALL YOU HEALTHY PEOPLE CANNOT IMAGINE THE ELATION WE EPILEPTICS FEEL DURING THE SECOND BEFORE OUR FIT.

In extreme cases, of whom the Russian novelist **Feodor Dostoyevsky** (1821-81) is the most celebrated, epileptics become ecstatic. They are filled by feelings of transcendence and beatitude, overwhelmed by the glory of existence.

Alternative Perceptions

Throughout history, and in all cultures, some epileptics and some consumers of psychedelics have held such experiences to be of supreme significance.

Aldous Huxley (1894-1963)
WE HAIL THEM AS REVELATION - THE DOORS TO AN ALTERNATIVE REALITY.
MODERN NEUROSCIENCES OFFER A VERY DIFFERENT FRAMEWORK OF EXPLANATION.

This framework refers only to the neurochemistry and electrophysiology of brain circuits. On the phenomenology and meaning of abnormal experience, as of normal experience, neuroscience remains silent.

Sanity: Beliefs and Pathologies

Several people convicted of witchcraft in the 17th century have descendants who suffer from **Huntington's disease**, the symptoms of which include twisting, twitching and grimacing. Throughout history, epileptics have also been accused of possession and suffered persecution.

HUMAN BEHAVIOUR IS ALWAYS INTERPRETED IN TERMS OF THE DOMINANT BELIEFS, OR NARRATIVES, OF THE CULTURE.

Religious societies give **supernatural** explanations of abnormal behaviour. Modern societies prefer a diagnosis of **medical pathology**, especially where there is a physical abnormality such as an epileptic seizure. However, where the abnormality is purely mental, as in **delusions**, there is considerable ambivalence still.

For example, is **schizophrenia** a disease of certain **dopamine** pathways in the brain (the medical model)? Or is it a mode of coping with intolerable personal circumstances (the phenomenological or sociological model)? It is not always clear that these are alternative rather than complementary forms of explanation.

Consider the visions of **Hildegarde of Bingen** (1098-1179), which she saw in an alert and wakeful state "with the eyes of the spirit and the inward ears".

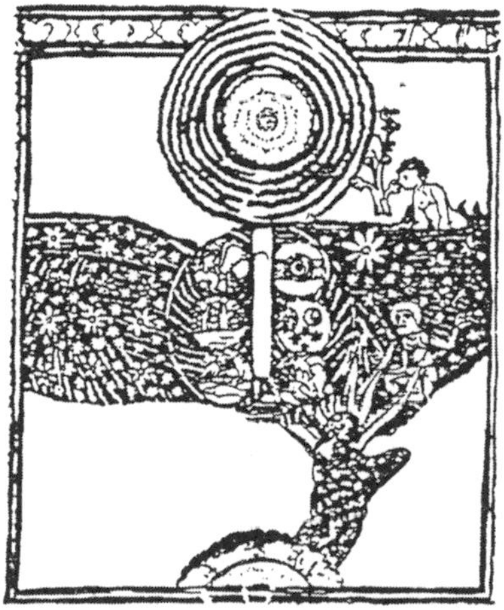

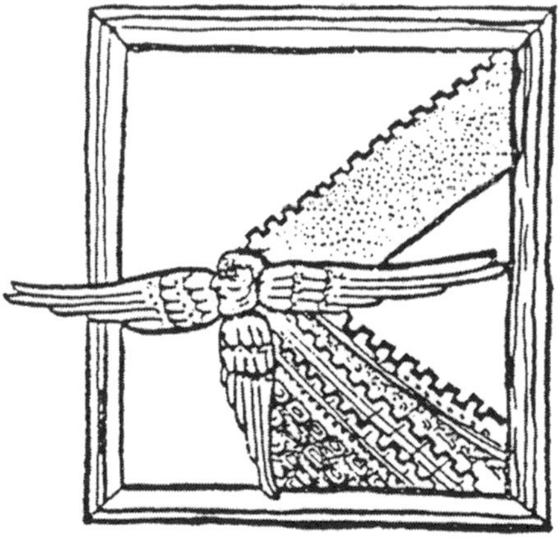

Typical fortification illusion of visual migraine

Hildegarde made detailed paintings of her visions, which she believed came from God. They exhibit the concentric circles, fortification-like figures and falling stars of what we now call **visual migraines**, themselves a kind of minor epilepsy.

Explaining Delusions

Neuroscience explains the *physical* basis of Hildegarde's visual disturbances. At the same time, we understand how a devout woman of the 12th century could sensibly arrive at a *spiritual* interpretation of them. **Cognitive neuropsychiatry** tries to show that delusional beliefs are attempts to explain pathological experiences. Let's begin with a "delusion" experienced in normal life.

Most of us have at some time sat on a train and been deluded into thinking we were moving, when it was the train next door that had started.

This mistake is understandable because it is usually only when **we** move that a large chunk of the environment slides across our retina.

Now let's look at the way that schizophrenics interpret their "voices".

Hearing Voices

In daily life we, or our brains, constantly distinguish sensory changes produced by our activity from changes produced by other people. We know when we have spoken or someone else has spoken. We recognize when someone else has given us an idea and when we have thought of one for ourselves.

In experiments involving a throat microphone and headphones, delusional schizophrenics sometimes reported that words they had said had been spoken by another person.

This supports the view that they experience their own speech and inner speech as "voices" and that their delusions are attempts to account for the disembodied speakers they hear.

The proposal is that schizophrenics have a brain defect that makes them bad at distinguishing their own silent speech (and thoughts) from external speech. In this respect, they remind us of Homer's Greeks hearing the commands of the gods.

The Impostors Delusion

Another example is the **Capgras** delusion. Capgras individuals may be generally quite lucid, yet they come to regard their parents, partners or children as "impostors", doubles pretending to be the people they resemble. Many Capgras cases have documented brain injury.

A recent idea is that this delusion may be a "mirror image" of **prosopagnosia** (see pp. 110-111). In prosopagnosia, it appears that conscious visual perception of faces happens normally but is **disconnected** from both (a) recognition of identity and (b) an emotional sense of facial recognition (see pp. 110-111).

The prosopagnosic **consciously** sees the man who is his father. There is also recognition of identity and emotional recognition, but these only occur **unconsciously**.

This is shown by the fact that prosopagnosics show bodily responses to familiar faces and learn true pairings of famous names and faces faster than fake pairings.

For the Capgras case, the suggestion is that conscious visual perception of faces happens normally, and recognition of identity reaches consciousness as normal, **but** the emotional sense of facial recognition does not happen either consciously or unconsciously. This person can see and identify his father but feels no emotional "glow" of recognition. The delusion that his father is an impostor may be the best way he can make sense of his lack of emotional reaction – less frightening than to accept that he has lost that capacity.

This person exhibits the Capgras delusion when he **sees** his parents, but not when he **hears** their voices on the telephone. He gives the same emotional reaction to familiar faces (including his parents) as to unfamiliar ones.

What Do We Learn About the Mind from Studying the Brain?

We can think of the brain as composed of numerous natural computers, each of which evolved to solve a particular problem by following its own set of rules (its algorithm). So, V1 and V2 respond to changes in the light on the retina. V3, V4 and V5 each takes a part of this information and computes shape, colour and motion, respectively. This information then feeds into areas in the temporal lobe that determine object and face recognition, and into areas in the parietal lobe that generate spatial representations. Each brain area is like a computer in a linked system. What it does only makes sense in the context of what the whole system does.

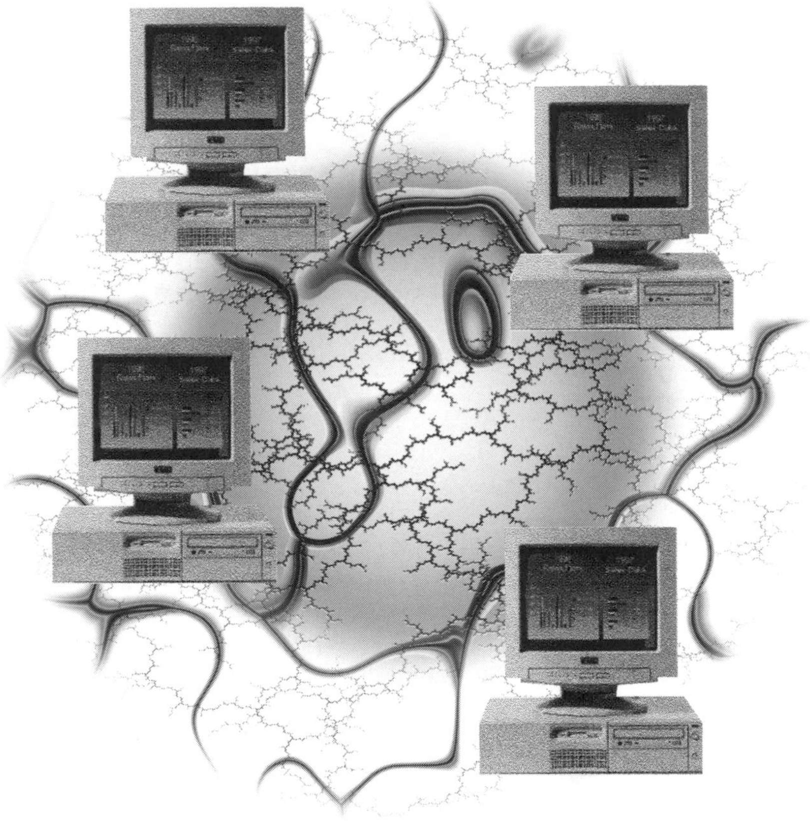

This is similar to the way in which the pumping action of the heart makes sense only in the context of a circulatory system for the blood.

Each brain area (or computer) can itself also be viewed as a system with component parts that act in concert to perform that area's role in the larger system. In the same way, the heart can be seen as a system of muscles, tubes, spaces and valves which act in concert to perform the pumping action that is the heart's role in the circulatory system.

Complex systems are nested within other complex systems. Finding the bottom layer of the hierarchy is impossible because you can always take the analysis one step further. For example, we have seen that terms like "vision" and "memory" prove to be very broad, covering many distinct processes and functions.

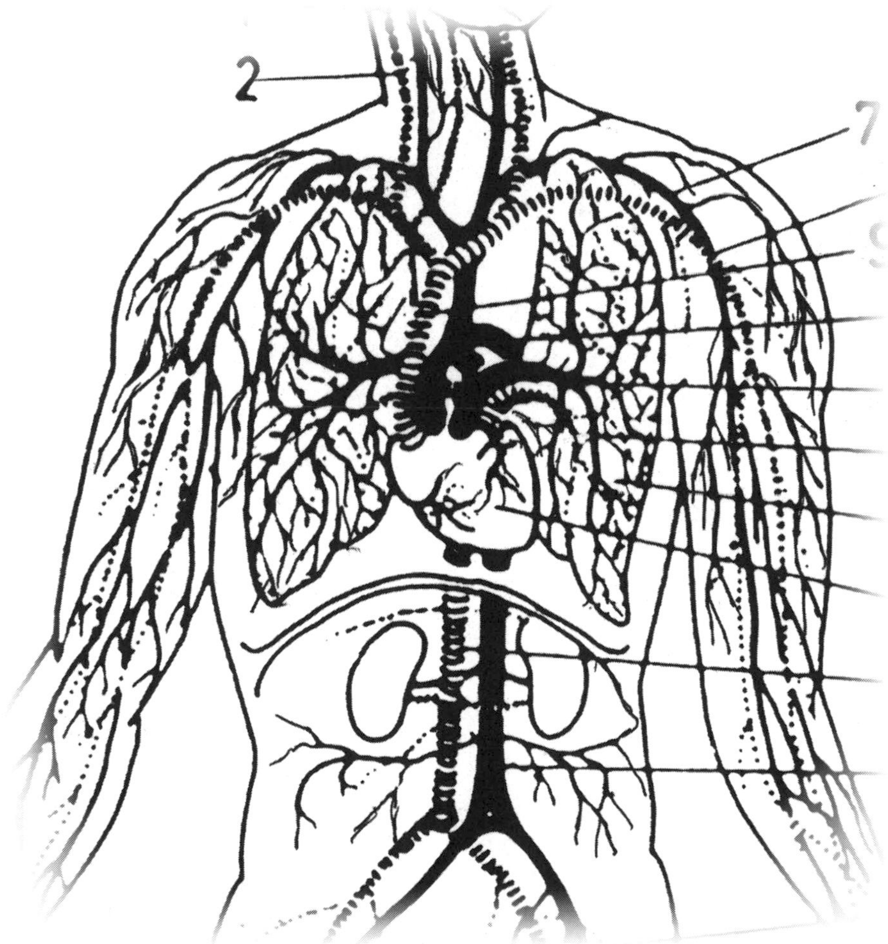

Evolution of the Mind

How did matters come to be this way? We assume that the mind evolved to solve the problems faced by primates in the wild.

Colour vision is useful for finding coloured fruits amongst green foliage.

And cognitive maps in memory are useful for finding the same fruit tree again next day or next year.

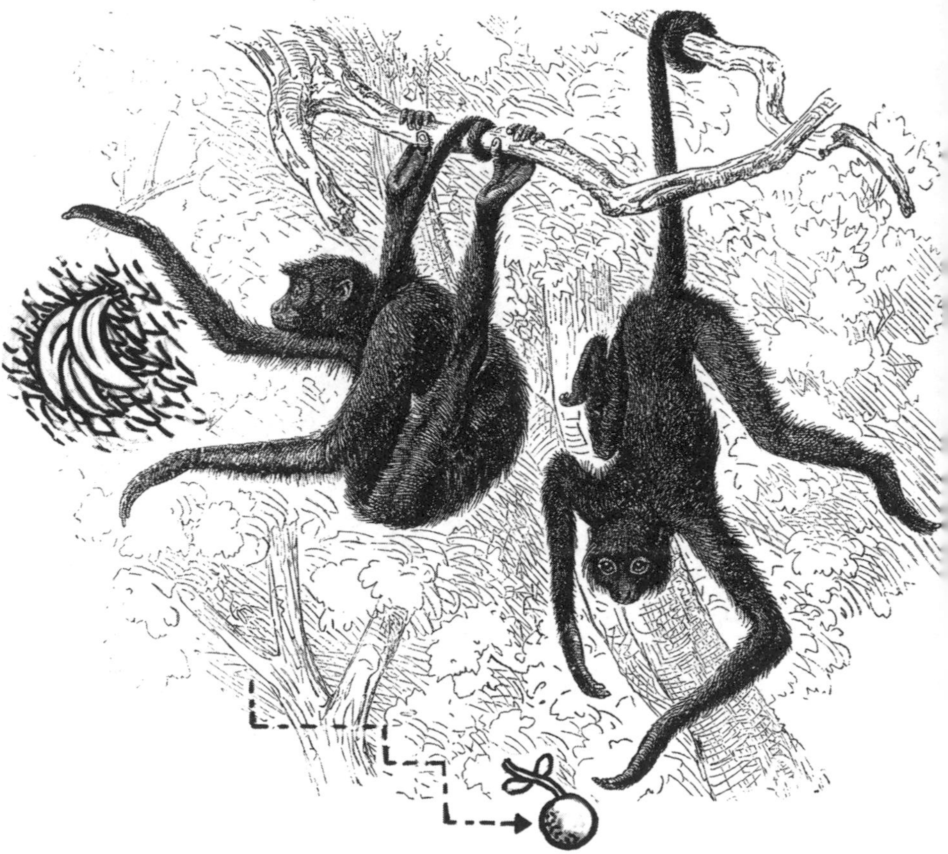

However, because they live in social groups, primates have a social environment to cope with as well as a physical environment. The social intellect hypothesis says that much of the evolution of the brain/mind may have been in response to the complexity of the *social* world rather than of the *physical* world.

The Social Intellect

Of course, being social does not guarantee evolution of a big brain. Ants make the point. However, ants appear not to recognize each other as individuals. One worker ant is as good as another because they all exhibit very similar, pre-wired behaviour. By contrast, animals who *learn* a lot of their behaviour are not so readily interchangeable.

Each may have habits but, being learnt, these differ between individuals. Therefore, the ability to recognize individuals becomes important, and a brain system for face recognition develops. For animals who recognize each other visually, it soon becomes worth knowing which individuals can and cannot be relied upon in this or that circumstance.

Humans are not the only ones to engage in this "social trade". To be effective at it, animals must not only recognize faces but also be able to predict individual behaviour. They have to be able to experience others as "personalities".

Mind Reading

Recently, it has been proposed that there is a "mind reading" module in the brain that allows us to experience a world of individuals with dispositions and preferences, just as our complex visual system allows us to experience a world of objects with particular shapes, colours, locations and movements. Mind reading is thought to involve the amygdala, the superior temporal sulcus, the medial frontal cortex and the orbito-frontal cortex.

If there is a mind reading module, then damage to it should produce abnormal experience of other minds, just as damage to the visual system gives abnormal visual experience.

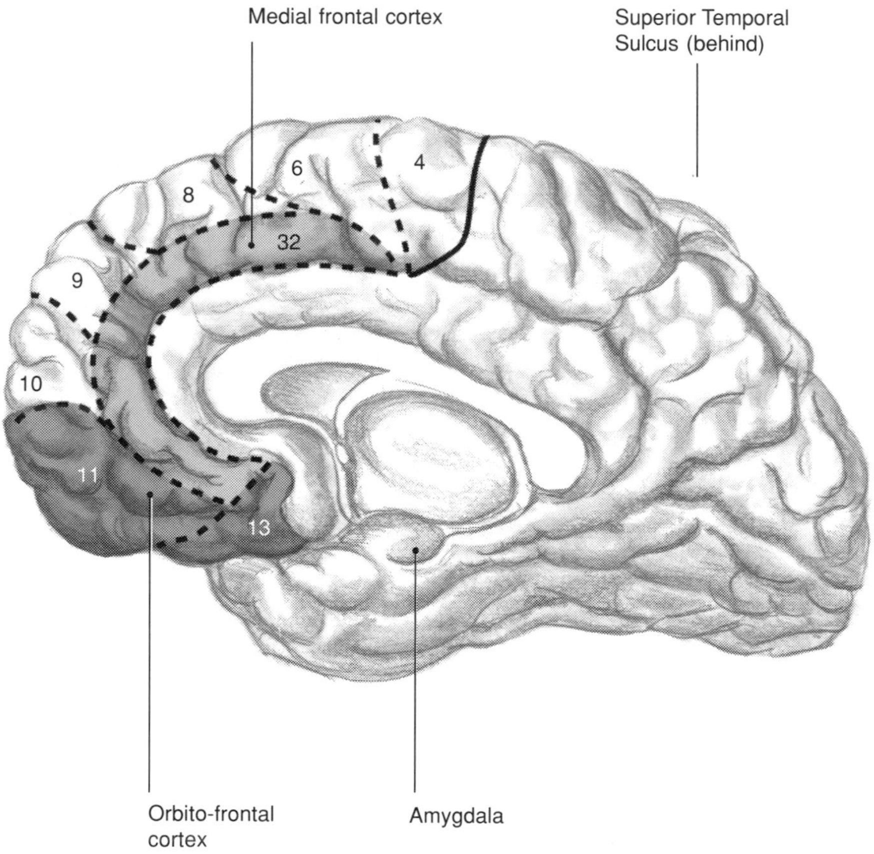

People with autism may have damage to this module. They seem to be "mind blind", unable to experience others as personalities with mental states.

Take this example of failure to understand mental states. An adult female shows a packet of sweets to an autistic boy.

Ordinary children and children with Down's syndrome pass this test with ease. Autistic children fail it. They do not seem to understand about other people's mental states.

Do Mental States Exist Outside Our Experience of Them?

If people can be mind blind to mental states, does that mean mental states do not exist outside our experience of them? Parallel questions can be asked about colour. Do colour blind people fail to detect colours that are out in the world waiting to be perceived? Or does colour blindness demonstrate that colour exists only in our conscious experience?

We can draw an analogy with "pain blind" individuals who lack the experience of pain and so frequently injure themselves. Nobody suggests there is pain out in the world and that such people are failing to detect it. Pain is **our** experience. When considered in this way, colours too seem to be **ours**.

Sight of a daffodil causes your sensation of yellow.

Just as the prick of a thorn causes your sensation of pain.

The Heider Experiment

On this argument, an encounter with another person causes your experience of their mental states. And just as the hunting response of the toad can be triggered by a matchstick moving lengthways, so our response of experiencing mental states can be triggered by any object superficially resembling a person.

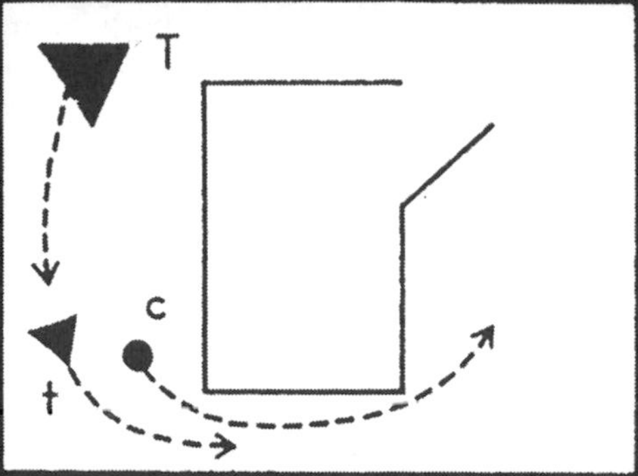

Almost anything that shows spontaneous movement or change will suffice. People project mental states and personality onto animals, the planets, rivers, volcanoes, the wind, the sea, cars, ships and – in one famous experiment – geometric shapes moving around a flat surface.

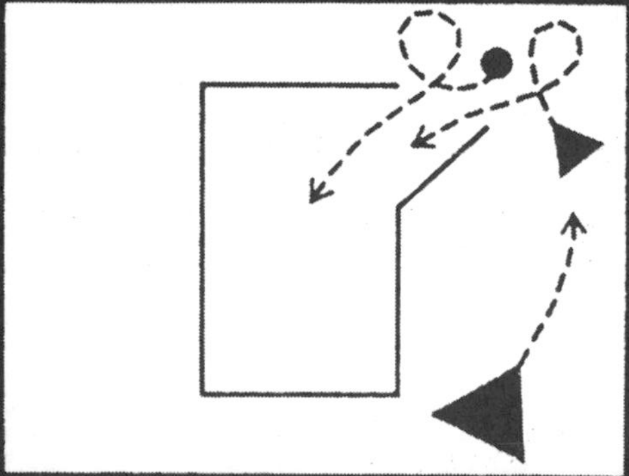

“The little triangle and the circle are scared of the big triangle. It chases them into the house and shuts the door to trap them.”

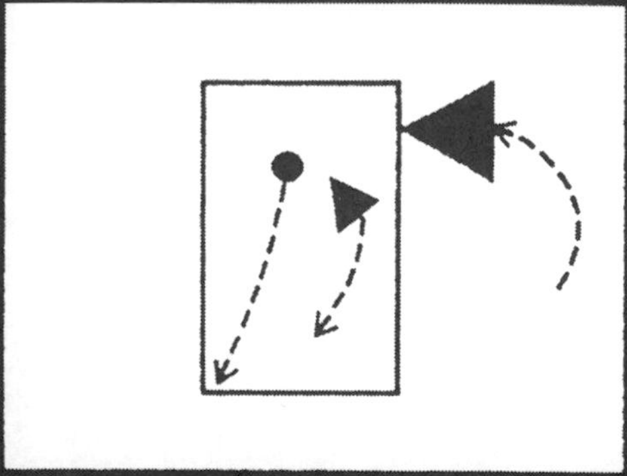

We have seen that vision and memory fractionate into many component processes. Other categories of our commonsense folk psychology have stood up no better to scrutiny. Emotion, attention, action and the self all fragment under inquiry. There is a multiplicity of selves. The narrative self is the most prominent. Yet the confabulations of people with brain injury show that the narrative self has limited understanding of the individual's behaviour. And now we are suggesting that people's mental states exist only in other people's experience of them.

This question must be answered with a resounding YES and NO.

What About Personal Responsibility?

If mental states exist only in other people's experience, and if the self is not a single moral agent but some conglomerate, what are the moral consequences? Surely, our culture claims to rest on a concept of personal moral responsibility.

Well, how did the Greeks cope with this issue?

Homer's characters excuse their more horrible actions, of which the epics contain many, on the grounds that they could not do otherwise. Injured parties accept such explanations and give similar accounts of their own actions. However, this does not stop them from wreaking vengeance. The Greeks reckoned you could be **answerable** for an act even though you were not **responsible** for it. This is not unlike the way parents may be legally answerable for the acts of their young children.

Homer's *Iliad* tells how King Agamemnon took the hostage Briseis away from Achilles.

Because evolution has equipped us all with very similar brains, people in all societies, including the ancient Greeks, "read" into behaviour what in our culture are termed intentions, desires and beliefs. For us, these are "mental states" which precede and cause behaviour. Apart from exceptional circumstances of diminished responsibility, we attribute these to the individual.

Other societies may read into behaviour dispositions rather than mental states. They may attribute these dispositions to the gods or to witchcraft, but without necessarily absolving the individual of liability for his or her actions.

Crime and Punishment

The circumstances in which a society punishes an individual are determined by interlocking practices in relation to personal responsibility, individual rights, the communal good, expediency, what are acceptable forms of punishment, and so on. In certain societies it is illegal to smack a child. In others, men have freedom to beat their wives and offspring. In still others, an absolute ruler may do as he wishes with his subjects.

Accepted practices vary. Yet every society reserves the right to protect its members from certain types of loss or injury by taking action against offenders.

Sometimes society imprisons (or even executes) a violent person, even though it is agreed that, due to insanity, he lacks responsibility for his actions. On other occasions, lack of responsibility can be used as a legal defence aimed at winning a **reduced** sentence, for example, when a defence of "provocation" or "crime of passion" is offered. Everyone knows how idiosyncratic judicial decisions can be.

We do not negotiate these difficult issues with any greater consistency or clarity of thought than did the Greeks.

But we do talk and think differently about such issues and, as a consequence, we lead our lives differently.

Study of the brain teaches us that human beings are complex in unsuspected ways. Behaviour arises from the co-action of many brain modules, and there is no single self exercising overall control. This does not mean the end of "morality as we know it". What it means is gradual transformation. "Morality as we know it" is a product of historical developments in how we think about personal responsibility, free will, rights, expediency and the good of the community.

In Britain, only two hundred years ago, a child could be hanged for stealing a sheep, and women did not enjoy equal political rights with men. Then there was the slave trade; now there is the arms trade.

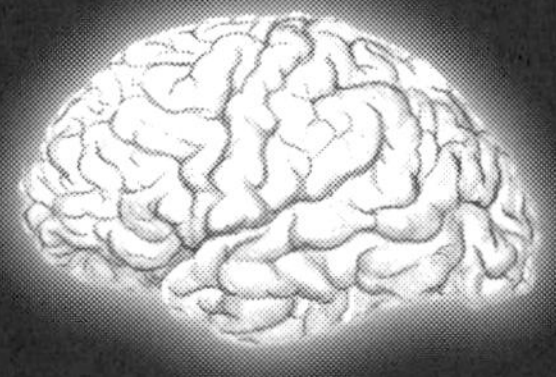

Further Reading

There are many books dealing with a greater or smaller part of the ideas covered in these pages. We can do no better than recommend some of those upon which we have relied ourselves.

History of neurosciences

The human brain and spinal cord: a historical study. E. Clarke and C.D. O'Malley. University of California Press, 1968. A broad and scholarly history of the development of knowledge and ideas about the brain.

Origins of neuroscience. S. Finger. Oxford University Press, 1994. A fascinating and excellently illustrated history of ideas.

The mind, the Greeks and literacy

The origins of European thought. R.B. Onians. Cambridge University Press, 1954. Authoritative analysis of the formative influence of Greek culture on the European intellect.

The origins of consciousness in the breakdown of the bicameral mind. J. Jaynes. Houghton Mifflin, 1976. Daring and thought-provoking interpretation of a number of early literatures, including the Homeric epics.

Brain and behaviour

The brain. Scientific American Library, 1979. Highly accessible but selective introduction to the structure and function of the brain.

Mind and brain. Scientific American Library, 1992. Further highly accessible but selective survey of current knowledge. Excellent illustrations.

Cognitive neuroscience: the biology of the mind. M.S. Gazzaniga, R.B. Ivry and G.R. Mangun. Norton & Co., 1998. Splendid, up-to-the-moment introduction to the whole topic by three leading practitioners.

A vision of the brain. S. Zeki. Blackwell Science, 1993. A renowned vision scientist presents an intriguing and personal account of one hundred years of study of the visual brain.

Human neuropsychology

The man who mistook his wife for a hat. O. Sacks. Duckworth & Co., 1985. Classic collection of case histories, written with deep humanity for a general readership.

Clinical neuropsychology. J.L. Bradshaw and J.B. Mattingley. Academic Press, 1995. Particularly well organized and clearly written introduction to the study of head-injured people.

Fundamentals of human neuropsychology. B. Kolb and I.Q. Whishaw. W.H. Freeman & Co., 1996. The comprehensive standard text for those wanting to discover what is known about the structure and function of the primate brain.

The Authors

Angus Gellatly is currently Head of the Department of Psychology at the University of Keele. He researches in visual perception and cognition, and used to write fiction when he could find the time.

Oscar Zarate has illustrated introductory guides to Freud, Stephen Hawking, Lenin, Mafia, Machiavelli, Quantum Theory and Melanie Klein. He has also produced many acclaimed graphic novels, including *A Small Killing*, which won the Will Eisner Prize for the best graphic novel of 1994, and has edited *It's Dark in London*, a collection of graphic stories, published in 1996.

Acknowledgements

The author would like to thank Melanie, Charlotte and Theo for their patience and support while he was immersed in this project. Thanks also to Melanie, Richard, Doug, Helen and Louise for reading and commenting on the original draft, and to Oscar for making it such an enjoyable collaboration.

The artist would like to thank Zoran Jevtic, who with the help of his mouse made an invaluable contribution to the visual effect of this book, particularly in the clarity of the maps of the brain.
I would also like to thank Angel Petronio Azarmendia, my friendly local librarian.

Credits

The photograph of Stephen Hawking on page 73 is by Mark McEvoy.
The drawing on page 125 is by Bill Elder.

Index